Springer Proceedings in Mathematics & Statistics

Volume 498

This book series features volumes composed of selected contributions from workshops and conferences in all areas of current research in mathematics and statistics, including data science, operations research and optimization. In addition to an overall evaluation of the interest, scientific quality, and timeliness of each proposal at the hands of the publisher, individual contributions are all refereed to the high quality standards of leading journals in the field. Thus, this series provides the research community with well-edited, authoritative reports on developments in the most exciting areas of mathematical and statistical research today.

Amine Laghrib · Mourad Nachaoui ·
Lekbir Afraites
Editors

Computational Methods for Inverse Problems and Applications

ICMDS 2024, Khouribga, Morocco, October 21–22

Springer

Editors
Amine Laghrib
Department of Mathematics
and Informatics
ENSA Khouribga
Sultan Moulay Slimane University
Khouribga, Morocco

Mourad Nachaoui
Department of Mathematics
FST Béni Mellal
Sultan Moulay Slimane University
Khouribga, Morocco

Lekbir Afraites
Department of Mathematics
FST Béni Mellal
Sultan Moulay Slimane University
Khouribga, Morocco

ISSN 2194-1009 ISSN 2194-1017 (electronic)
Springer Proceedings in Mathematics & Statistics
ISBN 978-3-031-89497-8 ISBN 978-3-031-89498-5 (eBook)
https://doi.org/10.1007/978-3-031-89498-5

Mathematics Subject Classification: 35R30, 68Q32, 34K29

This Springer imprint is published by the registered company Springer Nature Switzerland AG
The registered company address is: Gewerbestrasse 11, 6330 Cham, Switzerland

Preface

Mathematics continues to be a foundational element in the advancement of scientific and technological domains. Through its robust theoretical framework, mathematics provides essential tools for understanding complex phenomena in the natural and engineered worlds. Recent years have witnessed an unprecedented evolution in scientific methods and technological applications, driven by rapid progress in mathematical research and computational capabilities.

This special issue aims to capture the essence of these advancements, showcasing how mathematics has not only kept pace but has also been instrumental in shaping these transformative changes. It focuses on contemporary developments in applied mathematics and highlights the significance of these advancements in solving real-world problems.

The theme of this special issue is *Computational Methods for Inverse Problems and Applications*, focusing on cutting-edge research that intersects modern scientific needs. It features a collection of papers that cover diverse topics, especially in the application of mathematical tools to solve inverse problems. These problems are increasingly relevant due to advancements in technology and the growing complexity of real-world systems.

While the special issue respects traditional areas of applied mathematics, it also explores emergent themes that have gained significant attention in recent years. These include:

- Machine Learning and Deep Learning techniques applied to inverse problems.
- The interplay between Partial Differential Equations (PDEs) and Artificial Intelligence (AI).
- Applications of PDEs in Image Restoration.
- Mathematical modeling on Networks and Graphs.
- Mathematical frameworks for epidemiology, including COVID-19 models.
- Computational simulations related to climate change and environmental sciences.

This volume also reflects the growing trend of interdisciplinary research, where mathematics serves as a bridge between various fields, enriching both the theory and practical aspects of computational sciences. This collection comprises meticulously reviewed papers that were presented at the Third International Conference on Mathematical and Computational Sciences (ICMDS 2024), held in ENSA Khouribga, Morocco, from October 21 to 22, 2024. The conference brought together experts from around the globe to discuss recent advancements and emerging trends in applied mathematics and computational methods.

This special issue reflects the ongoing evolution of applied mathematics in the context of modern scientific and technological needs. It highlights how mathematics continues to play a pivotal role in the advancement of diverse fields by providing the computational tools necessary for solving some of the most complex and pressing problems of our time. With the rapid growth of machine learning, AI, and computational technologies, the contributions in this volume are expected to serve as a valuable resource for future research and applications.

Khouribga, Morocco

Amine Laghrib
Lekbir Afraites
Mourad Nachaoui

Contents

Solving a Time-Dependent Source Problem for a Time-Fractional Diffusion Equation with Two Numerical Approaches

Lekbir Afraites, Mohammed Srati, and Abdessamad Oulmelk

Abstract The aim of this study is to determine a time-dependent source problem in the time-fractional diffusion equation from the additional measurement data at an inner point. By transforming the inverse problem into an minimization problem, we employ two distinct alternative approaches to solve the latter control problem. To assess the effectiveness of these approaches, we test them under conditions with and without noise. The results obtained are very interesting and encouraging, indicating the potential success of the proposed methods.

Keywords Inverse source problem · Optimization problem · Fractional diffusion equation · ADAM and Landweber methods

1 Introduction

Time-fractional diffusion equations arise when using time fractional derivatives instead of the standard time derivatives. These equations serve as valuable tools for describing both sub-diffusion and super-diffusion phenomena [1–3]. In this context, analytical and numerical techniques have been used to study these equations [4–6].

However, in some practical situations, part of the boundary data, initial data, diffusion coefficient or source term may not given and we want to find them by additional measurement data. For the time-fractional diffusion equations, the inverse source problems have attracted immense interest in recent years. For this reason,

L. Afraites (✉) · M. Srati
EMI, FST Beni Mellal, Sultan Moulay Slimane University, Beni Mellal, Morocco
e-mail: l.afraites@usms.ma

A. Oulmelk
Laboratory of Mathematics and Applications, FST Tanger, Abdelmalek Essaadi University, Tangier, Morocco
e-mail: a.oulmelk@uae.ac.ma

A. Laghrib et al. (eds.), *Computational Methods for Inverse Problems and Applications*, Springer Proceedings in Mathematics & Statistics 498,
https://doi.org/10.1007/978-3-031-89498-5_1

a number of theoretical studies have examined in [7–10]. On the other hand, for numerical methods, several authors have attacked in [11–18].

This work is devoted to the study of an inverse problem involving a time-dependent source for a fractional diffusion equation in a bounded domain. For this, we consider the initial boundary value problem:

$$\begin{cases} {}^C D_t^s u(x,t) - \Delta u(x,t) + a(x)u(x,t) = \phi(t)\psi(x), & (x,t) \in Q_T := \Omega \times (0,T], \\ u(x,0) = f(x), & x \in \Omega, \\ u(x,t) = 0, & (x,t) \in \partial\Omega \times (0,T], \end{cases} \tag{1}$$

were Ω is a bounded domain in $\mathbb{R}$ with sufficiently smooth boundary $\partial\Omega$, $T > 0$ is the final time, ${}^C D_t^s u$ denotes the Caputo fractional left derivative of order $0 < s < 1$ with respect to t defined by

$${}^C D_t^s u(t) = \frac{1}{\Gamma(1-s)} \int_0^t \frac{u'(\tau)}{(t-\tau)^s} d\tau, \forall t > 0, \tag{2}$$

with $\Gamma(.)$ denotes Euler's Gamma function. If all functions ϕ, ψ, f, a are given appropriately, the problem (1) is a direct problem. The inverse problem here is to determine the source term $\phi(t)$ in problem (1) from the additional data

$$u(x_0, t) = h(t), \quad \forall t \in (0, T], \tag{3}$$

where $h(t)$ is a known function which satisfies the compatibility condition $f(x_0) = h(0)$, and $x_0 \in \Omega$ is an interior fixed point.

Numerous research works have been conducted to address the inverse source problem related to time-space fractional diffusion (1). For example, in [19], the authors successfully tackled an inverse problem involving a space-dependent source and also determined the orders of time and space fractional derivatives in a time-space fractional diffusion equation. Additionally, another study by the authors in [20] focused on solving an inverse problem using the Fourier truncation method. They provided convergence estimates and rules for choosing the regularization parameter but did not offer any numerical examples. In a different line of inquiry, [21] delved into investigating the Cauchy problem for non-linear pseudo-differential equations involving both time and space components. Meanwhile, addressing the inverse time-dependent source problem (1)–(3), researchers have conducted several studies. For instance, in [22], the uniqueness and stability estimate of this problem were thoroughly examined. In [7], researchers explored the application of a regularized minimization approach under two distinct cost functions to analyze an inverse source problem for a time-fractional diffusion equation. Moreover, [23] successfully tackled an inverse source problem for a subdiffusion equation utilizing the minimization method. For further insights into this inverse problem, comprehensive details can be found in [24, 25].

In the subsequent discussion, we delve into the inverse problem (1)–(3), which can be reformulated as an optimization task with the objective of solving the following minimization problem:

$$\bar{\phi} = \arg \min_{\phi \in \mathbb{D}_{ad}} \mathbb{J}(\phi), \tag{4}$$

where $\mathbb{J}$ is the cost function defined by

$$\mathbb{J}(\phi) = \frac{1}{2} \int_0^T |u(x_0, t; \phi) - h(t)|^2 dt + \frac{\lambda}{2} \int_0^T |\phi(t)| dt, \tag{5}$$

and λ is a regularization parameter and $\mathbb{D}_{ad}$ is the admissible set defined by as follows

$$\mathbb{D}_{ad} = \left\{ \phi \in L^2(I); \underline{b} \leq \phi(t) \leq \bar{b} \text{ a.e. } in\ I \right\}. \tag{6}$$

Within this context, we present two distinct numerical approaches aimed at resolving the control optimal problem (4). The initial method is known as ADAM, while the second is referred to as the Landweber method. Furthermore, we conduct a comprehensive comparative analysis between these two methods, examining their respective strengths and weaknesses.

The subsequent sections of this paper are structured as follows: In Sect. 2, we employ the ADAM method to address the optimization problem. Section 3 provides a comprehensive exposition of the Landweber method, which is utilized to numerically solve our inverse problem. Furthermore, Sect. 4 presents the numerical results obtained from each method and includes a comparative analysis of these results. Finally, the findings and key takeaways of this study are summarized in Sect. 5.

2 The ADAM Method

In this section, we propose Adam method which is an innovative technique for efficient stochastic optimization that exclusively relies on first-order gradients, minimizing memory usage. This method calculates adaptive learning rates individually for various parameters based on estimates of both first and second moments of the gradients. The name "Adam" stems from its foundation in adaptive moment estimation [26]. This method is designed to combine the advantages of two recently popular methods which are adaptive subgradient method [27] and Root Mean Squared Propagation (RMSprop) method [28]. We adapt this Adam algorithm here to numerically reconstruct the minimizer of the objective functional $\mathbb{J}$.

Unfortunately this functional is not differentiable because the regularization term is L^1. To overcome this problem, we approximate the regularization term by a regular function that is twice differentiable. In this context, this functional $\mathbb{J}$ can be approximated as follows:

$$\mathbb{J}^{(\epsilon)}(\phi) = \frac{1}{2} \int_0^T |u(x_0, t; \phi) - h(t)|^2 dt + \frac{\lambda}{2} \int_0^T |\phi^2(t) + \epsilon^2|^{\frac{1}{2}} dt, \tag{7}$$

where $\epsilon > 0$ is a positive parameter. Note that $\mathbb{J}^{(\epsilon)}(\phi)$ converges to $\mathbb{J}(\phi)$, when ϵ tends toward zero.

The approximate minimization problem then becomes

$$\bar{\phi} = \arg\min_{\phi \in \mathbb{D}_{ad}} \mathbb{J}^{(\epsilon)}(\phi), \tag{8}$$

In order to use the conjugate gradient method to solve the optimization problem (8), we need to calculate the gradient of the cost function $\mathbb{J}^{(\epsilon)}$, given by the following Theorem.

Theorem 1 *The gradient of the cost function $\mathbb{J}^{(\epsilon)}$ is given by the following expression*

$$\nabla \mathbb{J}^{(\epsilon)}(\phi) = \int_{\Omega} \psi(x) p(x, .) dx + \frac{\lambda \phi}{(\phi^2 + \epsilon^2)^{\frac{1}{2}}},$$

where p is the solution of following adjoint problem:

$$\begin{cases} {}^C D_t^s p(x,t) - \Delta p(x,t) + a(x) p(x,t) = F(x,t), & (x,t) \in Q_T, \\ p(x,0) = 0, & x \in \Omega, \\ p(x,t) = 0, & (x,t) \in \partial\Omega \times (0,T], \end{cases} \tag{9}$$

with $F(x,t) = (u(x,t) - h(t))\delta_{x_0}(x)$ and $\delta_{x_0}(.)$ is the Dirac function at point x_0.

Now, the Adam method to find the minimizer of $\mathbb{J}^{(\epsilon)}$ is given by Algorithm 1.

Algorithm 1 The Adam method for solving the nonlinear inverse problem.

```
Begin
Input: Choose an initial functions of iteration φ^(0), ψ^(0), ϑ^(0), x_0, tol, max_itr.
Initialize Set k = 0, β_1 = 0.9, β_2 = 0.999, ζ = 10^-7, ε = 10^-4, ρ = 0.001.
  while k < max_it do
    Solve the direct state (1) to obtain the solution u^(k) with φ = φ^(k).
    Solve the adjoint problem (9) to obtain p^(k) with φ^(k) = φ and F^(k) = u^(k)(x_0, t) − h(t).
    Compute ∇J^(ε)(φ^(k)) by using Theorem (1).
    Update first moment estimate by ψ^(k+1) = β_1.ψ^(k) + (1 − β_1)∇J^(ε)(φ^(k)).
    Update second moment estimate by ϑ^(k+1) = β_2.ϑ^(k) + (1 − β_2)(∇J^(ε)(φ^(k)))^2.
    Compute corrected first moment estimate by ψ̄^(k+1) = 1/(1−β_1^k) ψ^(k+1).
    Compute corrected second moment estimate by ϑ̄^(k+1) = 1/(1−β_2^k) ϑ^(k+1).
    Update φ^(k+1) = φ^(k) − ρ. ψ̄^(k+1) / (sqrt(ϑ̄^(k+1)) + ζ).
    Compute the errors E^(k) = ||φ^(k+1) − φ^(k)||_{L^2([0,T])} / ||φ^(k)||_{L^2([0,T])}
    Check convergence : if E^(k) ≤ tol stop.
    Set k = k + 1.
  end while
return φ^(k).
End
```

3 The Landweber Iteration Algorithm

In this section, we propose the second numerical method to solve the nonlinear inverse problem (1)–(3) by obtaining the numerical solution from the cost functional defined in (4). However, there are some deficiencies in this method. On one hand, we have to use the gradient method, introducing the solution of the adjoint problem (9), which differs from the standard problem and is difficult to solve. Additionally, finding the optimal regularization parameter λ presents a challenge. To address these issues, we utilize the Landweber iterative method to obtain the regularization solution for (1)–(3). Klann and Ramlau introduced this method [29] to tackle linear ill-posed problems. The main idea of the fractional Landweber method is based on iterative sequences, similar to the classical iterative method. Let $\mathcal{S}$ be the operator associated with the parameter to observation mapping, defined as follows

$$\mathcal{S} : \mathcal{D}_{ad} \subset L^2(0, T) \longrightarrow L^2(0, T)$$

defined by $\mathcal{S}(\phi)(t) = u(x_0, t; \phi) = h(t), \forall t \in (0, T]$. Hence, the inverse problem (1)–(3) can be rewritten in the following form

$$\text{Find } \phi \in \mathcal{D}_{ad} \text{ such that } \mathcal{S}\phi = h. \tag{10}$$

The nonlinear operator $\mathcal{S}$ can be divided into two operators $\mathcal{K}$ and $\mathcal{H}$ such that

$$\mathcal{S}(\phi) = \mathcal{K}(\phi) + \mathcal{H}(f) = u_1(x_0, t) + u_2(x_0, t), \tag{11}$$

where u_1 satisfies the following system

$$\begin{cases} {}^C D_t^s u_1(x, t) - \Delta u_1(x, t) + a(x) u_1(x, t) = \phi(t)\psi(x), & (x, t) \in Q_T, \\ u_1(x, 0) = 0, & x \in \Omega, \\ u_1(x, t) = 0, & (x, t) \in \partial\Omega \times (0, T], \end{cases} \tag{12}$$

and u_2 satisfies the following system

$$\begin{cases} {}^C D_t^s u_2(x, t) - \Delta u_2(x, t) + a(x) u_2(x, t) = 0, & (x, t) \in Q_T, \\ u_2(x, 0) = f(x), & x \in \Omega, \\ u_2(x, t) = 0, & (x, t) \in \partial\Omega \times (0, T]. \end{cases} \tag{13}$$

Consequently, the Landweber iteration method to deal with the operator Eq. (10) can be rewritten in the following form

$$\phi^{(k+1)} = (I - \mu\mathcal{K}^*\mathcal{K})\phi^{(k)} + \mu\mathcal{K}^*h, \tag{14}$$

where $\mu >$ and $\phi^{(0)}$ is an initial value of iteration for this landweber method.

From the definition of $\mathcal{K}$ and (14), we have

$$\begin{aligned}\phi^{(k+1)} &= \phi^{(k)} - \mu\mathcal{K}^*\big(\mathcal{K}(\phi^{(k)} - h\big)\\ &= \phi^{(k)} - \mu\mathcal{K}^*\big(u^{(k)}(x_0,.) - h\big),\end{aligned}$$

where $u^{(k)}$ is the solution of (12) with $\phi = \phi^{(k)}$.

From self-adjoint of $\mathcal{K}$, we can easily check that $\mathcal{K}^*(u(x_0.,t) - h(t)) = w(x_0.,t)$, where w solves the following problem

$$\begin{cases} {}^C D_t^s w(x,t) - \Delta w(x,t) + a(x)w(x,t) = \Psi(t), & (x,t) \in Q_T,\\ w(x,0) = 0, & x \in \Omega,\\ w(x,t) = 0, & (x,t) \in \partial\Omega \times (0,T], \end{cases} \tag{15}$$

where $\Psi(t) = u(x_0,t) - h(t)$. In addition, we obtain the Landweber method as expressed in Algorithm 2.

Algorithm 2 Landweber algorithm for solving the nonlinear inverse problem

Begin
Input: Choose an initial function of iteration $\phi^{(0)} \in \mathcal{D}_{ad}, h(t)$,$x_0, \mu, tol, max_{itr}$.
Initialize Set $k = 0$.
  while $k < max_{it}$ **do**
    Solve the problem (1) to obtain the solution $u^{(k)}$ where $\phi = \phi^{(k)}$.
    Solve the problem (15) to obtain $w^{(k)}$ where $\Psi(t) = u^{(k)}(x_0,t) - h(t)$.
    Update $\phi^{(k+1)}(t) = \phi^{(k)}(t) - \mu w^{(k)}(x_0,t)$.
    Compute the errors $E^{(k)} = \dfrac{\|\phi^{(k+1)} - \phi^{(k)}\|_{L^2([0,T])}}{\|\phi^{(k)}\|_{L^2([0,T])}}$
    Check convergence : **if** $E^{(k)} \le tol$ stop.
     Set $k = k + 1$.
  end while
return $\phi^{(k)}$.
End

4 Numerical Experiments

In this section, we present some numerical results to illustrate the performance of the proposed approaches. For this, we consider four examples in the one-dimensional case where the two first ones are smooth while the two last ones non-smooth. The additional data $h(t) = u(x_0,t)$ is obtained by solving the direct problem (1) by fixing all the unknown and known parameters of this problem as follows

$$a(x) = \sin(3\pi x) + 5, \quad \psi(x) = x, \quad f(x) = \sin(\pi x).$$

For solving the direct problem (1), we use the finite difference method.

To examine the numerical stability of our algorithms, we add noise to the data using the following formula

$$h^{\delta} = h + \delta.(2.rand(size(h) - 1),$$

where $rand$ is a random perturbation, and the magnitude δ is the noise level. To show the accuracy of the numerical solution, we calculate the approximate L^2 errors denoted by

$$e^{(1)} = \frac{\|\phi - \phi^{(k)}\|_{L^2([0,T])}}{\|\phi\|_{L^2([0,T])}},$$

where $\phi^{(k)}$ is the source reconstructed at the $k - th$ iteration, and ϕ is the exact solution. Next, in the numerical implementation,the final time T is always chosen as $T = 1.0$, the space domain Ω is taken as $\Omega = (0, 1)$, the internal fixed point $x_0 = 0.5$. In this case, the grid size for time and space variable in all four examples are fixed as $\delta t = \frac{1}{100}$, $\delta x = \frac{1}{150}$, respectively.

Example 1 In the first example, we suppose the unknowns smooth function is follows

$$\phi(t) = \sin(2\pi t) + \frac{5}{2}.$$

Example 2 In the second example, we consider the unknowns smooth functions is follows

$$\phi(t) = t^2 - t + \frac{1}{2}.$$

Example 3 In this example, we test a nonsmooth example. We suppose that the unknowns nonsmooth function is given by

$$\phi(t) = \begin{cases} 2t + \frac{7}{2}, & \text{if } 0 \leq t \leq \frac{1}{2}, \\ -2t + \frac{11}{2} & \text{if } \frac{1}{2} \leq t \leq 1, \end{cases}$$

Example 4 In the last example, we consider a discontinuous example. We suppose that the unknowns discontinuous function is follows

$$\phi(t) = \begin{cases} 4, & \text{if } 0 \leq t \leq \frac{1}{3}, \\ 5 & \text{if } \frac{1}{3} \leq t \leq \frac{2}{3}, \\ 4, & \text{if } \frac{2}{3} < t \leq 1. \end{cases}$$

4.1 Numerical Results Using the ADAM Method

In this section we give the numerical results for the previous examples using the first approach, the ADAM method without and with noise.

Reconstruction without noise This paragraph examines the results obtained from the first Algorithm 1 without noise. We begin by considering two examples of smooth functions: 1 and 2. Their numerical results are displayed in Fig. 1. Additionally, we analyze the results for two non-smooth examples: 3 and 4, with their corresponding numerical results shown in Fig. 2.

The advantage of this method is demonstrated by its successful application to smooth functions. However, the disadvantage lies in its inability to effectively capture singular points.

Reconstruction with noise This paragraph examines the results of the ADAM method in the presence of noise. To demonstrate the method's stability, we present

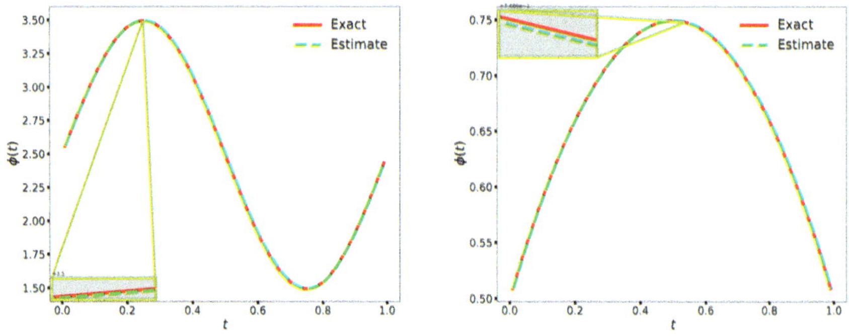

(a) Example 1 without noise (b) Example 2 without noise

Fig. 1 The numerical results of the ADAM method for Examples 1 and 2 without noise

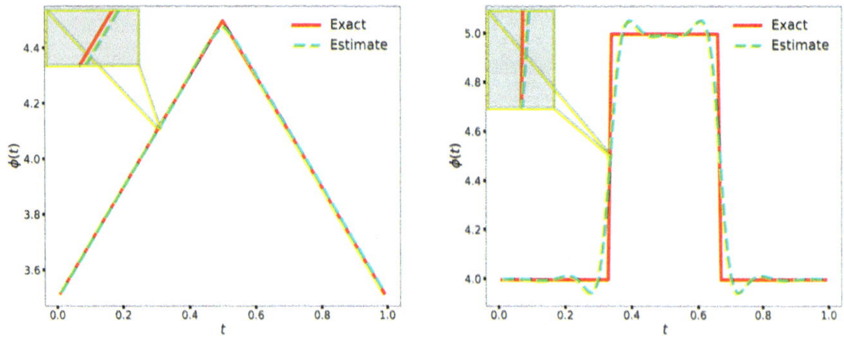

(a) Example 3 without noise (b) Example 4 without noise

Fig. 2 The numerical results of the ADAM method for Examples 3 and 4 without noise

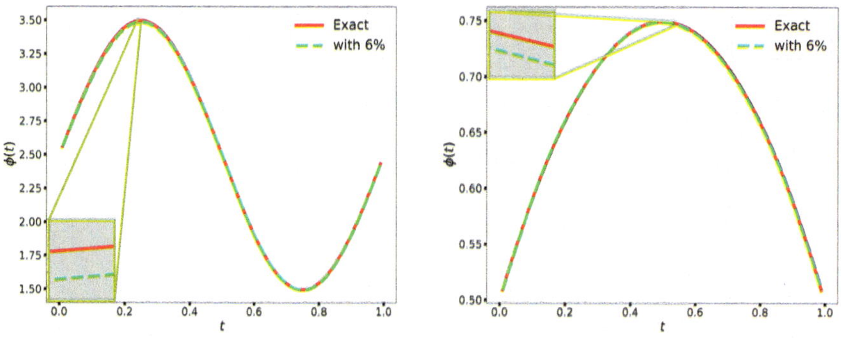

(a) Example 1 without noise (b) Example 2 without noise

Fig. 3 The numerical results of the ADAM method for Examples 1 and 2 with noise

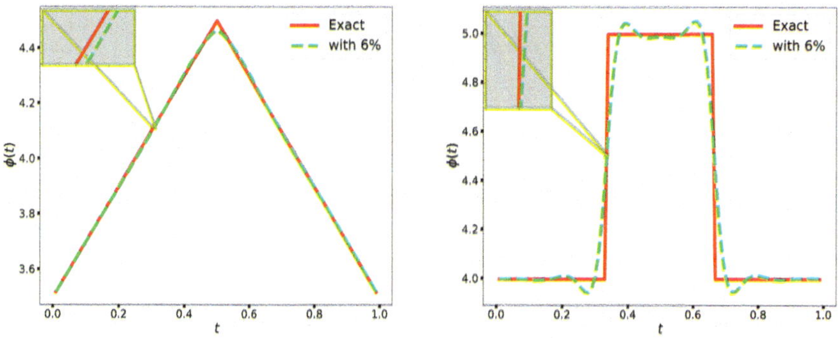

(a) Example 3 with noise (b) Example 4 without noise

Fig. 4 The numerical results of the ADAM method for Examples 3 and 4 with noise

numerical results for Examples 1 and 2 in Fig. 3. Additionally, we display the numerical results for Examples 3 and 4 in Fig. 4. Based on the results shown in Figs. 3 and 4, it can be concluded that this method exhibits numerical stability.

4.2 Numerical Results Using the Landweber Method

In this section, we present the numerical results for the previous examples using the Landweber method, both with and without noise

Reconstruction without noise We present the numerical tests for the second Algorithm (2) conducted without noise. The two smooth examples are illustrated in Fig. 5, while the non-smooth examples are displayed in Fig. 6. The figures showcase the

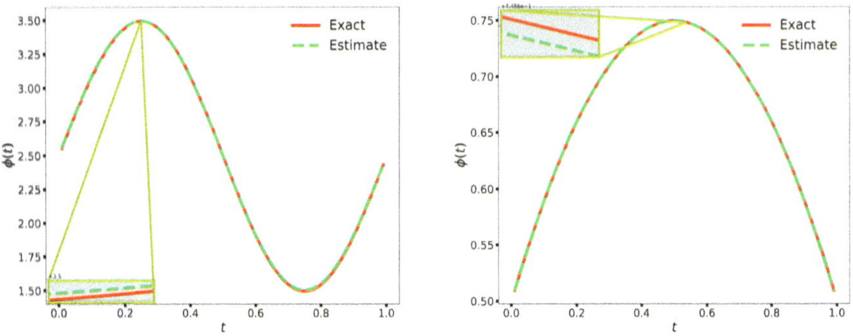

(a) Example 1 without noise

(b) Example 2 without noise

Fig. 5 The numerical results of the Landweber method for Examples 1 and 2without noise

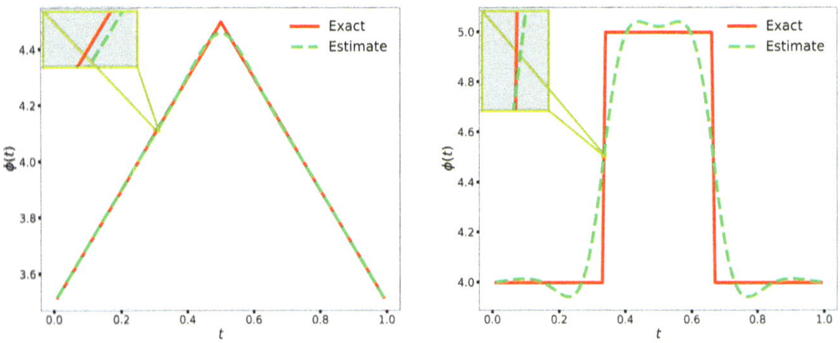

(a) Example 3 without noise

(b) Example 4 without noise

Fig. 6 The numerical results of the Landweber method for Examples 3 and 4 without noise

numerical results, indicating a perfect correspondence between the approximate solutions and the exact solution in the smooth test cases. However, in the non-smooth tests, the approximate solutions fail to capture the singular points accurately. These compelling results provide substantial evidence for the efficiency of the proposed method.

Reconstruction with noise In this part, we conducted numerical tests on the examples provided, introducing noise to the system. The results for the regular examples are illustrated in Fig. 7, while Fig. 8 displays the numerical outcomes for the singular examples. Based on these tests with noise, we can confidently conclude that our method exhibits numerical stability (Figs. 7 and 8).

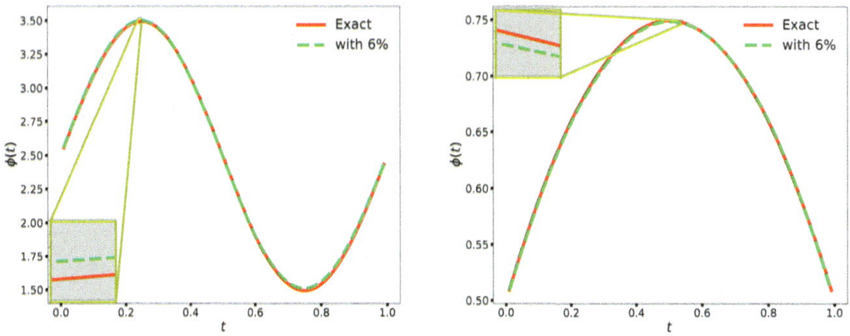

(a) Example 1 without noise

(b) Example 2 without noise

Fig. 7 The numerical results of the Landweber method for Examples 1 and 2 with noise

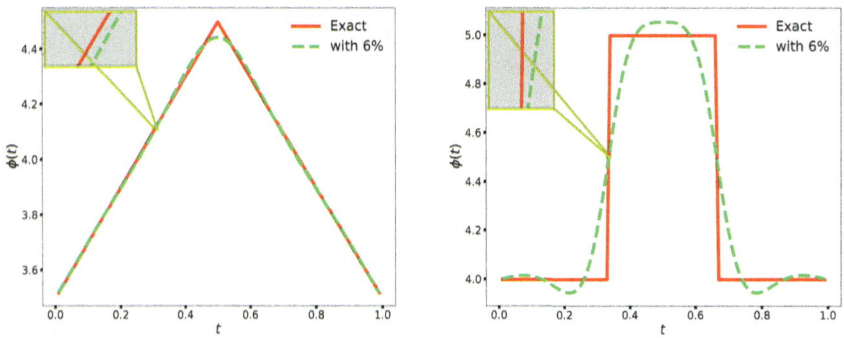

(a) Example 3 with noise

(b) Example 4 without noise

Fig. 8 The numerical results of the Landweber method for Examples 3 and 4 with noise

4.3 Comparison of the Proposed Methods

This paragraph compares the ADAM method with the Landweber method for all the aforementioned examples, based on reconstruction results and relative error. Figure 9 illustrates the numerical comparison between the ADAM and Landweber methods for Examples 1 and 2. Additionally, Fig. 10 presents the numerical comparison between the ADAM and Landweber methods for Examples 3 and 4. Furthermore, Fig. 11 displays the relative error for Examples 1 and 2 for both proposed algorithms, while Fig. 12 shows the relative error for Examples 3 and 4 for both proposed algorithms.

Based on the aforementioned results, it is evident that the ADAM method exhibits a higher level of acceleration compared to the Landweber method.

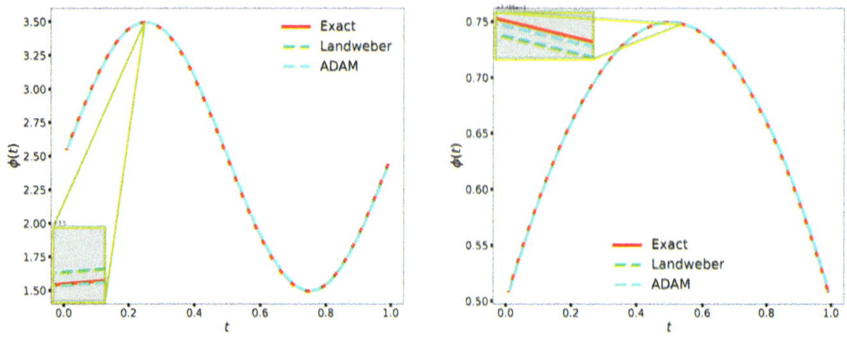

(a) Example 1 without noise (b) Example 2 without noise

Fig. 9 Numerical comparison between ADAM and Landweber methods for Examples 1 and 2

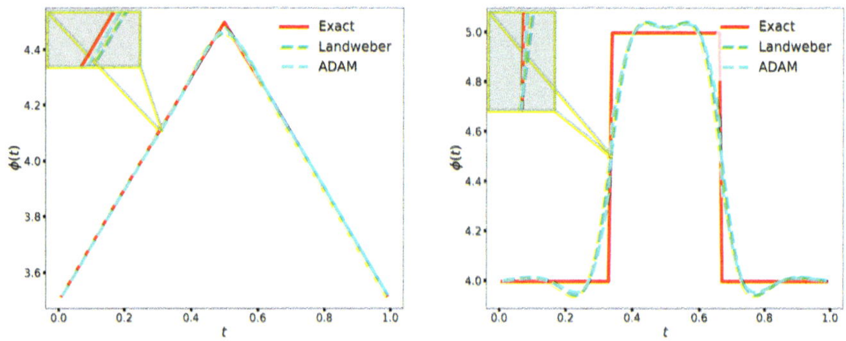

(a) Example 3 with noise (b) Example 4 without noise

Fig. 10 Numerical comparison between ADAM and Landweber methods for Examples 3 and 4

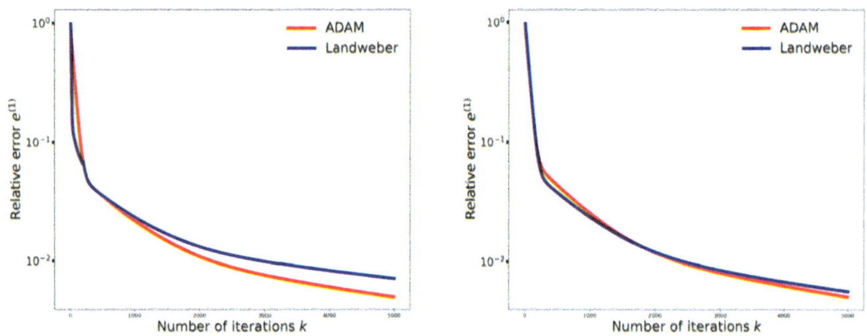

(a) Relative error for Example 1 (b) Relative error for Example2

Fig. 11 Relative errors of two different methods for Examples 1 and 2

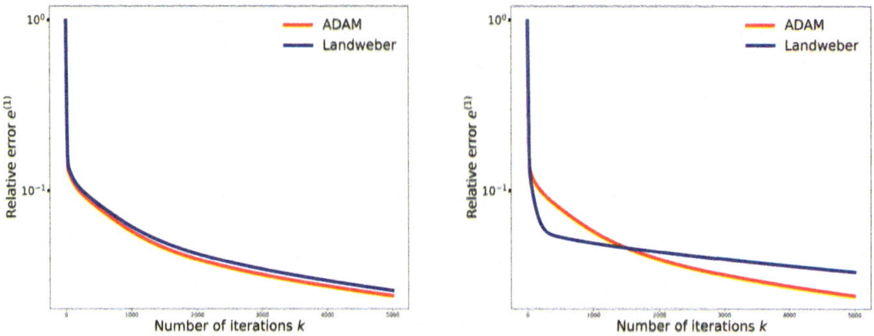

(a) Relative error for Example 3 (b) Relative error for Example 4

Fig. 12 Relative errors of two different methods for Examples 3 and 4

5 Conclusion

In this paper, we investigate a time-dependent source term in a time fractional diffusion equation. As part of this study, we propose two different alternative methods and compare them. Numerical examples indicate that both of the proposed methods are effective and stable. Furthermore, the ADAM method is found to be more effective than the Landweber method.

References

1. Berkowitz, B., Scher, H., Silliman, S.E.: Anomalous transport in laboratory-scale, heterogeneous porous media. Water Resour. Res. **36**(1), 149–158 (2000)
2. Metzler, R., Klafter, J.: The random walk's guide to anomalous diffusion: a fractional dynamics approach. Phys. Rep. **339**(1), 1–77 (2000)
3. Boukdir, A., Laghrib, A., Nachaoui, M.: Time-fractionated anisotropic osmosis with bilateral total variation for effective shadow mitigation in image processing. Appl. Math. Model. **137**, 115671 (2025)
4. Jin, B., Lazarov, R., Liu, Y., Zhou, Z.: The Galerkin finite element method for a multi-term time-fractional diffusion equation. J. Comput. Phys. **281**, 825–843 (2015)
5. Kian, Y., Oksanen, L., Soccorsi, E., Yamamoto, M.: Global uniqueness in an inverse problem for time fractional diffusion equations. J. Differ. Equ. **264**(2), 1146–1170 (2018)
6. Zaabouli, Z., El Hakoume, A., Afraites, L., Laghrib, A.: A novel coupled p (x) and fractional PDE denoising model with theoretical results. Int. J. Comput. Math. **101**(11), 1300–1325 (2024)
7. Oulmelk, A., Afraites, L., Hadri, A., Nachaoui, M.: An optimal control approach for determining the source term in fractional diffusion equation by different cost functionals. Appl. Numer. Math. **181**, 647–664 (2022)
8. Oulmelk, A., Srati, M., Afraites, L., Hadri, A.: An artificial neural network approach to identify the parameter in a nonlinear subdiffusion model. Commun. Nonlinear Sci. Numer. Simul. 107413 (2023)
9. Wei, T., Wang, J.-G.: Determination of robin coefficient in a fractional diffusion problem. Appl. Math. Model. **40**(17–18), 7948–7961 (2016)

10. Boukdir, A., Nachaoui, M., Laghrib, A.: Hybrid variable exponent model for image denoising: a nonstandard high-order PDE approach with local and nonlocal coupling. J. Math. Anal. Appl. **536**(2), 128245 (2024)
11. Afraites, L., Oulmelk, A.: Identification of robin coefficient in elliptic problem by a coupled complex boundary method. In: Mathematical Control and Numerical Applications: JANO13, Khouribga, Morocco, February 22–24, 2021, , pp. 71–86. Springer (2021)
12. Srati, M., Oulmelk, A., Afraites, L., Hadri, A., Zaky, M., Aldraiweesh, A., Hendy, A.: An inverse problem of determining the parameters in diffusion equations by using fractional physics-informed neural networks. Appl. Numer. Math. (2024)
13. Oulmelk, A., Afraites, L., Hadri, A., Zaky, M.A., Hendy, A., Zheng, X., Wang, H.: Nonsmooth optimization method for determining nonsmooth potential parameter in nonlinear subdiffusion equation. Commun. Nonlinear Sci. Numer. Simul. 108437 (2024)
14. Oulmelk, A., Srati, M., Afraites, L., Hadri, A.: Implementation of the ADMM approach to constrained optimal control problem with a nonlinear time-fractional diffusion equation. Discret. Contin. Dyn. Syst. S **18**(1), 15–42 (2025)
15. Srati, M., Oulmelk, A., Afraites, L., Hadri, A.: An ADMM approach to a TV model for identifying two coefficients in the time-fractional diffusion system. Fract. Calc. Appl. Anal. 1–36 (2023)
16. Jin, B., Zhou, Z.: Numerical estimation of a diffusion coefficient in subdiffusion. SIAM J. Control. Optim. **59**(2), 1466–1496 (2021)
17. Srati, M., Oulmelk, A., Afraites, L., Hadri, A.: An inverse problem of identifying two coefficients in a time-fractional reaction diffusion system. Discret. Contin. Dyn. Syst. S **18**(1), 113–147 (2025)
18. Oulmelk, A., Afraites, L., Hadri, A.: An inverse problem of identifying the coefficient in a nonlinear time-fractional diffusion equation. Comput. Appl. Math. **42**(1), 65 (2023)
19. Tatar, S., Tınaztepe, R., Ulusoy, S.: Simultaneous inversion for the exponents of the fractional time and space derivatives in the space-time fractional diffusion equation. Appl. Anal. **95**(1), 1–23 (2016)
20. Tuan, N.H., Long, L.D.: Fourier truncation method for an inverse source problem for space-time fractional diffusion equation. Appl. Anal. **91**(13), 2235–2248 (2017)
21. Kolokoltsov, V., Veretennikova, M.: Well-posedness and regularity of the Cauchy problem for nonlinear fractional in time and space equations (2014). arXiv:1402.6735
22. Li, Y., Wei, T.: An inverse time-dependent source problem for a time-space fractional diffusion equation. Appl. Math. Comput. **336**, 257–271 (2018)
23. Yan, X.B., Wei, T.: Inverse space-dependent source problem for a time-fractional diffusion equation by an adjoint problem approach. J. Inverse Ill Posed Probl. **27**(1), 1–16 (2019)
24. Zhang, Y., Xu, X.: Inverse source problem for a fractional diffusion equation. Inverse Probl. **27**(3), 035010 (2011)
25. Wang, W., Yamamoto, M., Han, B.: Numerical method in reproducing kernel space for an inverse source problem for the fractional diffusion equation. Inverse Probl. **29**(9), 095009 (2013)
26. Kingma, D.P., Ba, J.: Adam: a method for stochastic optimization (2014). arXiv:1412.6980
27. Duchi, J., Hazan, E., Singer, Y.: Adaptive subgradient methods for online learning and stochastic optimization. J. Mach. Learn. Res. **12**(7), (2011)
28. Hinton, G.E.: A practical guide to training restricted Boltzmann machines. In: Neural Networks: Tricks of the Trade, 2nd edn, pp. 599–619. Springer (2012)
29. Klann, E., Ramlau, R.: Regularization by fractional filter methods and data smoothing. Inverse Probl. **24**(2), 025018 (2008)

The Game p-Laplacian Evolution Equation on Graphs: From Image Processing to Machine Learning Applications

Yassine Belkheiri, Abdelwahhab Boudjelal, Abderrahim Elmoataz, and Sophie Schüpp

Abstract Inspired by recent research on the normalized p-Laplacian operator on graphs, we investigate the Cauchy problem associated with these operators. We establish the existence and uniqueness of solutions to diffusion problems and explore applications in image processing, 3D point cloud processing, and data classification.

Keywords Game p-Laplacian · Graph signal processing · Dirichlet problem · Image processing

1 Introduction

The game-theoretic p-Laplacian, has been recently introduced in the context of the stochastic game model known as Tug-of-War with noise [5, 8, 9]. This operator is particularly noteworthy as it encompasses specific cases of the p-Laplace depending on the value of p. For instance, it includes the mean curvature operator when $p = 1$, a scalar multiple of the standard Laplace operator when $p = 2$, and the ∞-Laplace when $p = \infty$. In homogeneous situations, the equation of game p-Laplacian matches the equation of variational p-Laplacian. This latter equation has been approximated through various schemes, such as finite element methods [1] and finite difference approximation schemes [7]. Additionally, a semi-Lagrangian scheme-based approx-

Y. Belkheiri · A. Boudjelal (✉) · A. Elmoataz · S. Schüpp
GREYC Laboratory, Normandie Univ, UNICAEN, CNRS, Caen, France
e-mail: abdelwahhab.boudjelal@unicaen.fr

Y. Belkheiri
e-mail: yassine.belkheiri@unicaen.fr

A. Elmoataz
e-mail: abderrahim.elmoataz@unicaen.fr

S. Schüpp
e-mail: sophie.schuepp@unicaen.fr

A. Boudjelal
Electrical Engineering Laboratory LGE, M'sila University, M'sila, Algeria

A. Laghrib et al. (eds.), *Computational Methods for Inverse Problems and Applications*, Springer Proceedings in Mathematics & Statistics 498,
https://doi.org/10.1007/978-3-031-89498-5_2

imation was suggested in [2, 4, 11], employing statistical operations for the application of the p-Laplacian. Nevertheless, all these methods assume regular domains. Future and existing applications necessitate addressing this issue in more general domains or graphs with arbitrary topology.

Motivated by the need to broaden the application of this operator to diverse discrete domains, Elmoataz et al. [3] introduced an adapted and generalized form of the normalized p-Laplacian. This novel approach establishes a new class of p-Laplacian on graphs, effectively bridging the nonlocal 1-Laplacian, nonlocal ∞-Laplacian, and nonlocal 2-Laplacian on graphs.

The present study investigates the Cauchy problem associated with this operator on graphs, establishing the existence and uniqueness of a solution for this diffusion problem. Expanding on the definition of the normalized p-Laplacian presented by Elmoataz et al. (2017), we revisit key definitions related to the normalized p-Laplacian on graphs, utilizing statistical operators. Lastly, we highlight the versatility of this operator by demonstrating its applications in diverse fields such as image processing, data processing, and machine learning.

1.1 Game P-Laplacian

The variational p-Laplacian of a function $f : \Omega \subset \mathbb{R}^N \to \mathbb{R}$ is defined for $1 \leqslant p < \infty$ as follows:

$$\Delta_p f = \operatorname{div}\left(|\nabla f|^{p-2} \cdot \nabla f\right). \tag{1}$$

For $p = \infty$, the traditional definition is:

$$\Delta_\infty f = \sum_{i,j}^{n} \frac{\partial f}{\partial x_i} \frac{\partial f}{\partial x_j} \frac{\partial^2 f}{\partial x_i \partial x_j}.$$

The game-theoretic p-Laplacian, also known as the normalized p-Laplacian, was presented in [7] to represent the stochastic Tug-of-War game. For $1 \leq p < \infty$, it is expressed as:

$$\Delta_p^N f = \frac{1}{p} |\nabla f|^{2-p} . \operatorname{div}\left(|\nabla f|^{p-2} \cdot \nabla f\right). \tag{2}$$

When $p = \infty$, the normalized p-Laplacian is given by:

$$\Delta_\infty^N f = |\nabla f|^{-2} \Delta_\infty f. \tag{3}$$

The operator $\Delta_p^N f$ is termed "normalized" due to its homogeneity of degree 1, meaning that $\Delta_p^N(tf) = t \cdot \Delta_p^N f$ for any $t \in \mathbb{R}$. This differs from the variational

p-Laplacian, which has homogeneity of degree $p-1$. Consequently, parabolic problems that involve the normalized p-Laplacian exhibit scaling invariance, a beneficial property particularly in image processing applications. For a smooth function f, Eq. (2) expressed as follows:

$$\begin{aligned} \Delta_p^N f &= \frac{(p-2)}{p}\Delta_\infty^N f + \frac{1}{p}\Delta f \\ &= \frac{(p-2)}{p}\Delta_\infty^N f + \frac{2}{p}\Delta_2^N f \\ &= \alpha(p)\Delta_\infty^N f + \beta(p)\Delta_2^N f, \end{aligned} \tag{4}$$

with $\alpha(p) = (p-2)/p$ and $\beta(p) = 2/p$.

For $p = 1$, the game p-Laplacian can be expressed as:

$$\Delta_1^N f = \operatorname{div}\left(\frac{\nabla(f)}{|\nabla(f)|}\right)|\nabla(f)|. \tag{5}$$

As $\Delta_1^N f = \Delta - \Delta_\infty^N f$, Eq. (4) can be rewritten as:

$$\Delta_p^N f = \alpha'(p)\Delta_2^N f + \beta'(p)\Delta_1^N f, \tag{6}$$

with $\alpha'(p) = \frac{2(p-1)}{p}$ and $\beta'(p) = \frac{2-p}{p}$.

It can be noted that to guarantee the positivity of both the α and β functions in Eq. (4), the range of p is restricted to $2 \leq p \leq \infty$. Likewise, for Eq. (6), the interval for positive values of α' and β' is limited to $1 \leq p \leq 2$.

With this in mind, the game p-Laplacian for the entire range of $1 \leq p \leq \infty$ can be expressed asfollows:

$$\Delta_p^N f = \begin{cases} \frac{2}{p}\Delta_2^N f + \frac{p-2}{p}\Delta_\infty^N f & \text{for } 2 \leq p \leq \infty \\ \frac{2(p-1)}{p}\Delta_2^N f + \frac{2-p}{p}\Delta_1^N f & \text{for } 1 \leq p \leq 2. \end{cases} \tag{7}$$

2 Game p-Laplacian on Graph

This section focuses on the game p-Laplacian on weighted graphs. We start by introducing statistical operators and considering a weighted graph $G = (V, E)$ and weighting function w, which represents the weights or similarities associated with the graph. These elements are essential for constructing the proposed normalized p-Laplacian. Following this, we scrutinize the existence and uniqueness of solutions for the homogeneous normalized p-Laplacian equation on graphs, taking Dirichlet boundary conditions into account.

2.1 Nonlocal Statistical Operators on Weighted Graphs

In order to establish the normalized p-Laplacian on graphs, we begin by introducing the following averaging operators, which generalize the traditional average operators (Mean, Max, Min, Midrange, Median):

$$\begin{aligned}
\text{NLMean}(f)(u) &= \frac{\sum_{v\sim u} w(u,v)(f(v))}{\sum_{v\sim u} w(u,v)}, \\
\text{NLMax}(f)(u) &= \max_{v\sim u}(\sqrt{w(u,v)}\max(f(v)-f(u),0)) + f(u), \\
\text{NLMean}(f)(u) &= \max_{v\sim u}(\sqrt{w(u,v)}\max(f(u)-f(v),0)) + f(u), \\
\text{NLMidrange}(f)(u) &= \frac{1}{2}(NLMin(f)(u) + N\,\text{LMax}(f)(u)), \\
\text{NLMedian}(f)(u) &= \text{median}\left((\nabla_w f)(u)\right) + f(u),
\end{aligned} \tag{8}$$

The median is defined using the classical discrete median operator as follows:

$$\text{median}_{1\leq j\leq m}\left\{x_j\right\} = \begin{cases} y_{\frac{m+1}{2}} & \text{if } m \text{ is odd} \\ \frac{y_{\frac{m}{2}}+y_{\frac{m}{2}+1}}{2} & \text{if } m \text{ is even} \end{cases} \tag{9}$$

where $x_i \in \mathbb{R}$ for $i = 1\ldots,m$.

With $\{y_1,\ldots,y_m\}$ being a non-decreasing rearrangement of $\{x_1,\ldots,x_m\}$, it is apparent that when we set $w(u,v) = 1$, we recover the classical statistical Mean, Median, and Midrange filters.

Definition 1 *We introduce the following definitions for the normalized versions of the nonlocal* 1*-Laplacian,* 2*-Laplacian, and* ∞*-Laplacian:*

$$\begin{aligned}
\Delta^N_{w,2}(f)(u) &= \text{NLMean}(f)(u) - f(u), \\
\Delta^N_{w,1}(f)(u) &= \text{NLMedian}(f)(u) - f(u), \\
\Delta^N_{w,\infty}(f)(u) &= \text{NLMidrange}(f)(u) - f(u).
\end{aligned} \tag{10}$$

We can observe that these operators are indeed related to partial operators on graphs.

$$\begin{aligned}
\Delta^N_{w,2}(f)(u) &= \frac{1}{\mu(u)}\Delta_{w,2}(f)(u), \\
\Delta^N_{w,\infty}(f)(u) &= \Delta_{w,\infty}(f)(u), \\
\text{and } \Delta^N_{w,1}(f)(u) &= \text{median}(\Delta_w(f)(u)).
\end{aligned} \tag{11}$$

2.2 Game P-Laplacian on Graph

By discretizing Eq. (7) and employing the discrete counterparts of the game p-Laplacian for $p = 1$, $p = 2$ and $p = \infty$,, we suggest a nonlocal formulation of the game p-Laplacian on graphs. This is expressed through the following equations:

$$\Delta^N_{w,p}(f)(u) = \begin{cases} \frac{2}{p}\Delta^N_{w,2}(f)(u) + \frac{p-2}{p}\Delta^N_{w,\infty}(f)(u) & \text{for } 2 \leq p \leq \infty, \\ \frac{2(p-1)}{p}\Delta^N_{w,2}(f)(u) + \frac{2-p}{p}\Delta^N_{w,1}(f)(u) & \text{for } 1 \leq p \leq 2. \end{cases} \tag{12}$$

Using the Definitions (10) and (12), the formulation of the game p-Laplacian on weighted graphs can be rewritten as:

$$\Delta^N_{w,p}(f)(u) = \text{NLA}(f)(u) - f(u) \tag{13}$$

where $\text{NLA}(f)(u)$ is a nonlocal average operator defined by

$$\text{NLA}(f)(u) = \begin{cases} \frac{2}{p}\text{NLMean}(f)(u) + \frac{p-2}{p}\text{NLMidrange}(f)(u), & \text{for } 2 \leq p \leq \infty, \\ \frac{2(p-1)}{p}\text{NLMean}(f)(u) + \frac{2-p}{p}\text{NLMedian}(f)(u), & \text{for } 1 \leq p \leq 2. \end{cases} \tag{14}$$

3 Existence and Uniqueness

Given a function $g : V \to \mathbb{R}$ we study the following Cauchy problem

$$\begin{cases} \frac{\partial}{\partial t} f(x,t) = \Delta^N_{w,p} f(x,t) & \text{in } V \times [0,T], \\ f(x,0) = g(x) & \text{in } V. \end{cases} \tag{P^G_N}$$

$$\Delta^N_{p,w}(f)(u) = NLA(f)(x,t) - f(x,t),$$

where $NLA(f)(x,t)$ is the nonlocal average operator defined in [[3], Eq. (3.7)]. The operator $\Delta^N_{p,w}(f)(u)$ can be rewritten in the following form

$$\begin{aligned} \Delta^N_{p,w}(f)(x,t) = \Delta_F(f)(x,t) &= F(f((y_1 \sim x),t),\ldots,f((y_m \sim x),t)) - f(x,t) \\ &= F(f((y \sim x),t)) - f(x,t), \quad y \in V_m(x), \end{aligned}$$

where the operator F will be described later on. We can see that f is a solution of (P_N^G) only if it is a solution of the integral equation

$$f(x,t) = K_g f(x,t), \tag{15}$$

where

$$K_g f(x,t) := \int_0^t e^{s-t} F(f((y_1 \sim x),t),\ldots,f((y_m \sim x),t))\,\mathrm{d}s + e^{-t}g(x).$$

We verify that the operators defined in [[3], Eq. (3.1)] are all averaging operators according to the following definition taken from [12].

Definition 2 *Let $F : \mathbb{R}^m \to \mathbb{R}$ be a continuous function. We call F an averaging operator if it satisfies the following set of conditions:*

i. $F(0,\ldots,0) = 0$ *and* $F(1,\ldots,1) = 1$;
ii. $F(tx_1,\ldots,tx_m) = tF(x_1,\ldots,x_m)$ *for all* $t \in \mathbb{R}$
iii. $F(t+x_1,\ldots,t+x_m) = t + F(x_1,\ldots,x_m)$ *for all* $t \in \mathbb{R}$
iv. *F is nondecreasing with respect to each variable.*

Remark 1 It hold's that, if $(x_1,\ldots,x_m),(y_1,\ldots,y_m) \in \mathbb{R}^m$, then

$$x_j \le y_j + \max_{1\le j\le m}\{x_j - y_j\} \text{ for all } j \in \{1,\ldots,m\}.$$

Let F be an averaging operator. As a result of combining (iii) and (iv), we have

$$\mathrm{F}(x_1,\ldots,x_m) \le \mathrm{F}(y_1,\ldots,y_m) + \max_{1\le j\le m}\{x_j - y_j\}.$$

Therefore

$$\mathrm{F}(x_1,\ldots,x_m) - \mathrm{F}(y_1,\ldots,y_m) \le \max_{1\le j\le m}\{|x_j - y_j|\}.$$

and moreover

$$|\mathrm{F}(x_1,\cdots,x_m) - \mathrm{F}(y_1,\ldots,y_m)| \le \max_{1\le j\le m}\{|x_j - y_j|\}.$$

Lemma 1 *The operators defined in [[3], Eq. (3.1)] are averaging operators.*

Proof – $NLMean(f)(x,t) = F_1(f((y_1 \sim x),t),\ldots,f((y_m \sim x),t)) = \frac{1}{\sum_{y\sim x} w(x,y)} \sum_{y\sim x} w(x,y)f(y,t).$

i. $F_1(0,\ldots,0) = 0$ and $F_1(1,\ldots,1) = \frac{1}{\sum_{y\sim x} w(x,y)} \sum_{y\sim x} w(x,y) = 1$

ii. For all $s \in \mathbb{R}$,

$$\begin{aligned}F_1\left(sf\left((y_1 \sim x), t\right), \ldots, sf\left((y_m \sim x), t\right)\right) &= \frac{1}{\sum_{y\sim x} w(x,y)} \sum_{y\sim x} w(x,y) s f(y,t) \\ &= \frac{s}{\sum_{y\sim x} w(x,y)} \sum_{y\sim x} w(x,y) f(y,t) \\ &= sF_1\left(f\left((y_1 \sim x), t\right), \ldots, f\left((y_m \sim x), t\right)\right).\end{aligned}$$

iii. For all $s \in \mathbb{R}$

$$\begin{aligned}&F_1\left(s + f\left((y_1 \sim x), t\right), \ldots, s + f\left((y_m \sim x), t\right)\right) \\ &= \frac{1}{\sum_{y\sim x} w(x,y)} \sum_{y\sim x} w(x,y)(s + f(y,t)) \\ &= \frac{1}{\sum_{y\sim x} w(x,y)} \left(s \sum_{y\sim x} w(x,y) + \sum_{y\sim x} w(x,y) f(y,t)\right) \\ &= s + \frac{1}{\sum_{y\sim x} w(x,y)} \sum_{y\sim x} w(x,y) f(y,t) \\ &= s + F_1\left(f\left((y_1 \sim x), t\right), \ldots, f\left((y_m \sim x), t\right)\right).\end{aligned}$$

iv. F_1 is nondecreasing with respect to each variable. Indeed, for $i \in \{1, \ldots, m\}$, taking $g : V \times [0, T] \to \mathbb{R}$, such that $g\left((y_i \sim x), t\right) \geq f\left((y_i \sim x), t\right)$. Since the weight function w is positive, we have

$$F_1\left(g\left(y_i, t\right)\right) \geq F_1\left(f\left(y_i, t\right)\right).$$

We prove existence and uniqueness of a solution using a fixed point argument for K_g, and in the same way, we prove for $NLMax$, $NLMin$, $NLMidrange$, and $NLMedian$.

Theorem 1 *Let $g \in L^\infty(V, \mathbb{R})$. Then there exists a unique solution f in $L^\infty(V, \mathbb{R})$ of (P_N^G).*

Proof *Existence.* Let $T > 0$ and

$$L_T^\infty := \left\{f \in L^\infty(V \times [0, T], \mathbb{R}) : f(x, t) \quad \text{is continuous in } t\right\}.$$

One can easily check that L_T^∞ is a Banach space with the L^∞-norm. By construction, the operator F is a linear combination between averaging operators, then, it is an averaging operator. Therefrom, we can see that K_g is a contraction on L_T^∞. In fact, using Remark 1, we have that

$$\left\| K_g f_1 - K_g f_2 \right\|_\infty \leq \int_0^t e^{s-t} ds \, \|f_1 - f_2\|_{L^\infty} \leq \left(1 - e^{-T}\right) \|f_1 - f_2\|_\infty ,$$

for all $f_1, f_2 \in L_T^\infty$. Therefore, by the Brouwer fixed-point theorem, K_g has a unique fixed point $f \in L_T^\infty$.

Uniqueness. Let f_1, f_2 be two solutions of (P_N^G) such that

$$f_1, f_2 \in C\left(0, T; L^\infty(V, \mathbb{R})\right).$$

Then, f_1, f_2 are solutions of (15) and therefore they are fixed points of K_g. Thus $f_1 \equiv f_2$ in $V \times [0, T]$ for all $T > 0$ due to K_g is a contraction operator.

4 Experiments

4.1 Numerical Experiments

In this section, we show the discrete Cauchy problem associated with p-Laplacian operator. For this, we solve the discrete evolution Cauchy problem (P_N^G) for which the initial function g is application-dependent.

To iteratively solve (P_N^G), we employ an explicit forward Euler time discretization:"

$$\frac{\partial f}{\partial t}(u,t) = \frac{f^{n+1}(u) - f^n(u)}{\Delta t}, \tag{16}$$

where $f^n(u) = f(u, n\Delta t)$.

Hence, we can try to solve (P_N^G) by the following iteration scheme:

$$\begin{cases} f^{n+1}(u) = f^n(u) + \Delta t \Delta_{w,p}^N (f^n)(u) & u \in V \\ f^0(u) = f_0(u) & u \in V. \end{cases} \tag{17}$$

Using $\Delta_{w,p}^N = \text{NLA}(f) - f$ and setting $\Delta t = 1$, we get the nonlocal average filter

$$\left\{ f^{n+1}(u) = \begin{cases} \frac{2}{p} \text{NLMean}(f^n)(u) + \frac{p-2}{p} \text{NLMidrange}(f^n)(u) & \text{for } 2 \leqslant p \leqslant \infty \\ \frac{2(p-1)}{p} \text{NLMean}(f^n)(u) + \frac{2-p}{p} \text{NLMedian}(f^n)(u) & \text{for } 1 \leqslant p \leqslant 2. \end{cases} \right. \tag{18}$$

We get simple non-local average operators.

4.2 Weighted Graph Construction

There are several prevalent methodologies for mapping discrete data $\{u_1, \ldots, u_n\}$ onto a weighted graph G. Given a vertex set $V(G)$, these graphs are constructed by defining edges E that encode neighborhood relationships, leveraging a pairwise distance function $d : V(G) \times V(G) \to \mathbb{R}^+$.

In the context of images, graphs structured around geometric neighborhoods are particularly effective in representing the spatial configuration and the function defined over this space. Notable methodologies include:

- **Grid Graphs**: These graphs are the most intuitive for representing images, where each pixel is connected to its adjacent pixels. Common grid graphs include those with 4-adjacency and 8-adjacency; larger adjacency schemes can be utilized to capture nonlocal neighborhood relationships.
- **k-Nearest Neighbor (k-NN) Graphs**: In these graphs, each vertex is connected to its k nearest neighbors based on the distance metric d. This generally forms a directed graph due to the asymmetry in neighborhood relationships. An undirected graph can be obtained by introducing an edge between vertices u and v if u is among the k nearest neighbors of v, or if v is among the k nearest neighbors of u.

The edge weights quantify the similarity between vertices such that

$$w(u, v) = \begin{cases} s(u, v) & \text{if } (u, v) \in E(G) \\ 0, & \text{otherwise} \end{cases}$$

where $E(G) \to \mathbb{R}^+$ is a similarity function. Typically, one can choose:

- $s_0(u, v) = 1$;
- $s_1(u, v) = e^{-\frac{d(u,v)}{\sigma}}$ with $\sigma > 0$, where d is a metric controlling the similarity between edges, and σ is a scale parameter.
- For patch-based methods, the similarity function is

$$s_2(u, v) = e^{-\frac{d(u,v)^2}{\sigma^2}},$$

where $d(u, v) = \|P(v) - P(u)\|_2$, and $P : u \in V \mapsto P(u) \in \mathbb{R}^m$ is the patch extraction operator at u. For each node/vertex u, $P(u)$ is an m-dimensional real vector containing, e.g., spatial coordinates, intensities, etc., of the neighbors of u. This definition of patches is valid only for grid graphs and cannot be considered for arbitrary graphs. To compute the patch on a 3D point cloud, the reader is referred to [6].

Where $E(G) \to \mathbb{R}^+$ denotes a similarity function. Common choices include:

- $s_0(u, v) = 1$;
- $s_1(u, v) = e^{-\frac{d(u,v)}{\sigma}}$, where $\sigma > 0$ is a scale parameter and d is a metric defining the similarity between edges.
- For patch-based approaches, the similarity function is given by

$$s_2(u, v) = e^{-\frac{d(u,v)^2}{\sigma^2}},$$

where $d(u, v) = \|P(v) - P(u)\|_2$, and $P : u \in V \mapsto P(u) \in \mathbb{R}^m$ is the patch extraction operator at u. For each vertex u, $P(u)$ is an m-dimensional vector that may include spatial coordinates, intensities, etc., of u's neighbors. This patch definition is specific to grid graphs and is not applicable to arbitrary graphs. For information on computing patches on a 3D point cloud, refer to [6].

4.3 *Results*

Filtering Fig. 1 demonstrates the filtering effects on an image, while Fig. 2 illustrates the effects on a colored 3D point cloud. The weight functions, denoted as w_n, are computed from the colors of the images or the point cloud. The results are displayed for three scenarios: $w_n = s_0$ (constant weight), $w_n = s_1$ (color-based), and $w_n = s_2$ (patch-based).

In cases where $w_n \neq 1$, an adaptive filtering process is employed, taking into account the differences in colors within the image or point cloud. This adaptive approach enhances the preservation of specific features of the graph signal, dependent on the graph weights. Notably, when using patch-based weights ($w_n = s_2$), repetitive or textured patterns are better preserved, while still providing the expected simplification effects.

Semi-supervidsed classification The previously introduced game p-Laplacian approach can be adapted for semi-supervised classification through discrete label diffusion by game p-Laplacian on graph. Let $G = (V, E, w)$ be a weighted graph, where data points are represented as vertices connected by edges in E. Semi-supervised clustering of V involves partitioning the set V into k classes (previously known), based on initial labels assigned to certain vertices in V. The objective is to then predict labels for the unlabeled data using the information from the labeled ones. Let C_l be a set of labeled vertices belonging to the lth class. Define $V_0 = \bigcup\{C_l\}_{l=1,\ldots,k}$, representing the set of initially *labeled* vertices, and let $V \setminus V_0$ be the initially unlabeled vertices. Each vertex $v_i \in V$ is characterized by a label vector, denoted as $f^0(v_i) = (f_l^0(v_i))^T_{l=1,\ldots,k}$ with

$$f_l^0(v_i) = \begin{cases} +1 & \text{if } v_i \in C_l \\ -1 & \text{if } v_i \in C_m | m \neq l \\ 0 & \forall v_i \in V \setminus V_0 \end{cases}$$

Fig. 1 Colored Image Filtering with the Normalized p-Laplacian $\boldsymbol{\Delta}^{\mathrm{N}}_{w_n,p}$. First column: local 8-adjacency grid graph ($w_n = 1$). Second column: same, with w_n = color. Last column: nonlocal with larger neighborhood (w_n depending on patches). Results for $p = 1$, $p = 2$, $p = 10$, and $p = \infty$

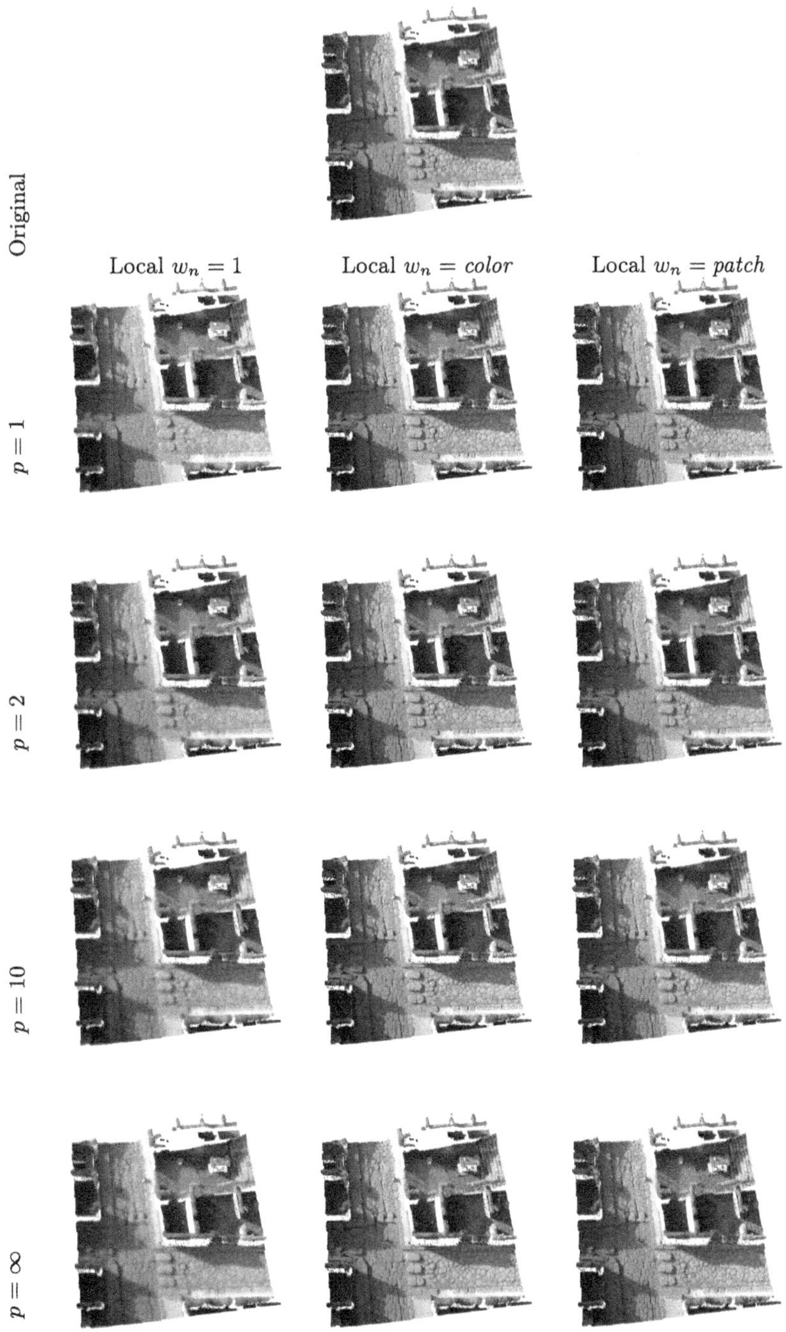

Fig. 2 Colored point cloud filtering with the Normalized p-Laplacian $\boldsymbol{\Delta}^{\mathrm{N}}_{w_n,p}$. First column: local 8-adjacency grid graph ($w_n = 1$). Second column: same, with w_n = color. Last column: nonlocal with larger neighborhood (w_n depending on patches). Results for $p = 1$, $p = 2$, $p = 10$, and $p = \infty$

Subsequently, the labeling of vertices involves conducting k independent game p-Laplacian diffusion processes to estimate membership functions $f_l : V \to \mathbb{R}$ for each class l.

Following the completion of the label propagation processes, class membership probabilities are estimated. The final classification for a given vertex $v_i \in V$ can then be obtained through $\arg\max_{l \in 1,\ldots,k} \left\{ \frac{f_l^{(t)}(v_i)}{\sum_l f_l^{(t)}(v_i)} \right\}$.

Real Data Clustering

MNIST, Optdigits, and Pendigits Datasets For the MNIST dataset (70,000 elements), the model consistently demonstrates high performance across different seed numbers (see Table 1 and Fig. 3), with mean scores ranging from 0.963 to 0.977. Notably, a marginal improvement is observed when p is set to 2.0 compared to ∞ and 1.0, suggesting potential benefits from a moderate penalty on coefficient magnitudes.

Turning to the Optdigits dataset (5,620 elements), the model's performance exhibits considerable variability (see Table 2 and Fig. 4), with mean scores ranging from 0.636 to 0.982. The sensitivity to the choice of the regularization parameter is evident, particularly with a substantial enhancement in performance when p is set to 2.0.

Likewise, for the Pendigits dataset (10,992 elements), the model consistently performs well (see Table 3 and Fig. 5), achieving mean scores ranging from 0.910 to 0.987. A modest improvement in performance is observed when p is set to 2.0, emphasizing the influence of the regularization parameter.

In general, the observed variability in results across datasets suggests that the impact of regularization and seed number may be contingent upon dataset characteristics. Given this sensitivity, further exploration of a broader range of p values

Table 1 Results for MNIST dataset (70000 elements) with different p values and seed numbers

Seeds (%)	p (Parameter value)	Mean score
1	∞	0.963774
1	1.0	0.963449
1	2.0	0.970171
2	∞	0.96869
2	1.0	0.969916
2	2.0	0.970696
5	∞	0.973554
5	1.0	0.97274
5	2.0	0.971116
10	∞	0.977306
10	1.0	0.974261
10	2.0	0.97138

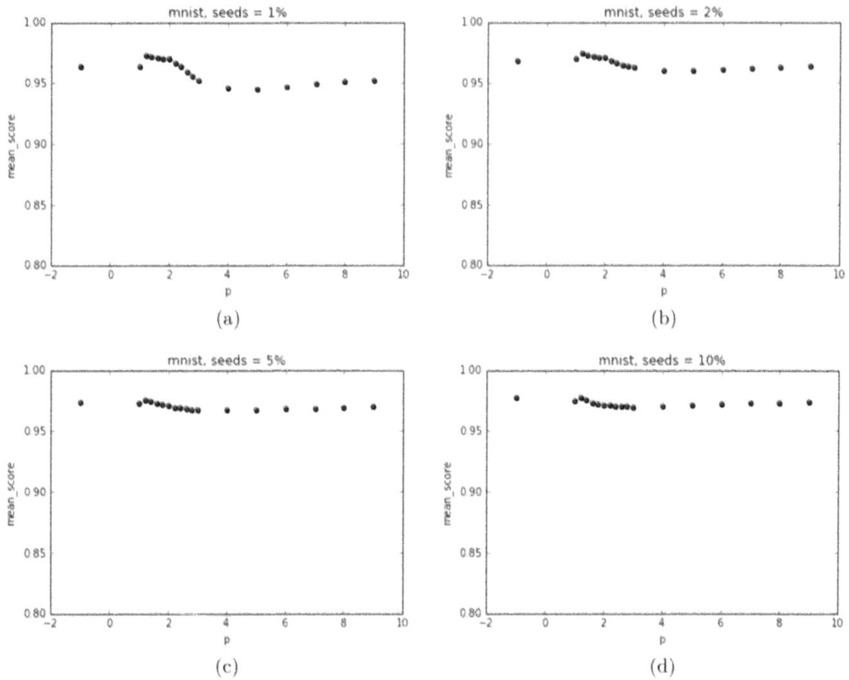

Fig. 3 Overall Caption for the MNIST Figures

Table 2 Results for optdigits dataset (5620 elements) with different p values and seed numbers

Seeds (%)	p (Parameter value)	Mean score
1	∞	0.948363
1	1.0	0.636103
1	2.0	0.803221
2	∞	0.967527
2	1.0	0.654662
2	2.0	0.823523
5	∞	0.977064
5	1.0	0.75653
5	2.0	0.82347
10	∞	0.982669
10	1.0	0.810534
10	2.0	0.835783

and seeds is advisable. Additionally, future research could delve into the model's sensitivity to additional hyperparameters and employ statistical tests to ascertain the significance of observed performance differences.

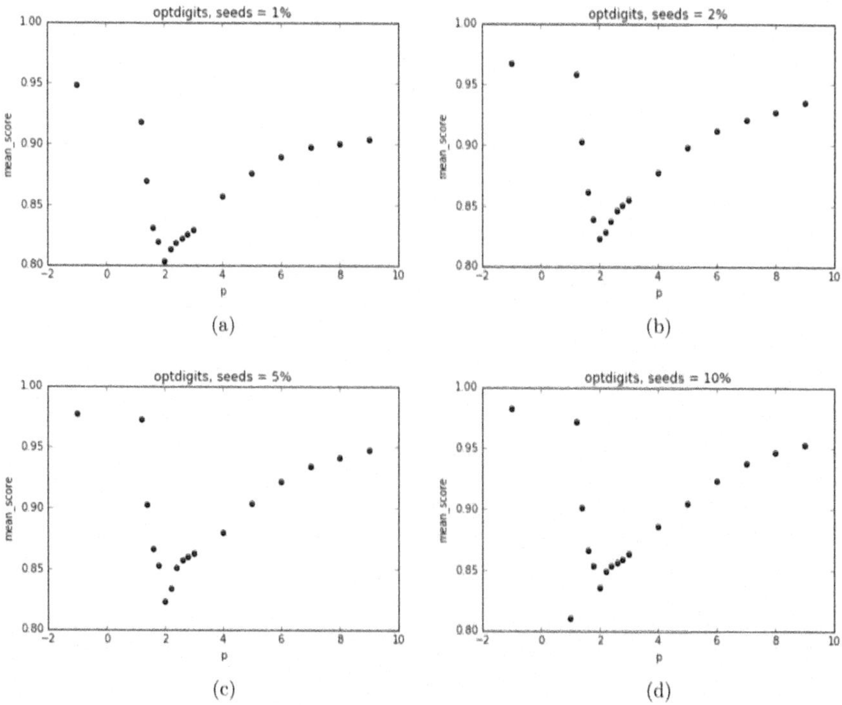

Fig. 4 Overall Caption for the Optdigits Figures

Table 3 Results for pendigits dataset (10992 elements) with different p values and seed numbers

Seeds (%)	p (Parameter value)	Mean score
1	∞	0.94506
1	1.0	0.910535
1	2.0	0.934298
2	∞	0.964693
2	1.0	0.913746
2	2.0	0.954039
5	∞	0.979421
5	1.0	0.945024
5	2.0	0.964511
10	∞	0.987464
10	1.0	0.963592
10	2.0	0.973344

Now, we turn our attention to the presentation of experiments focusing on semi-supervised classification of data.

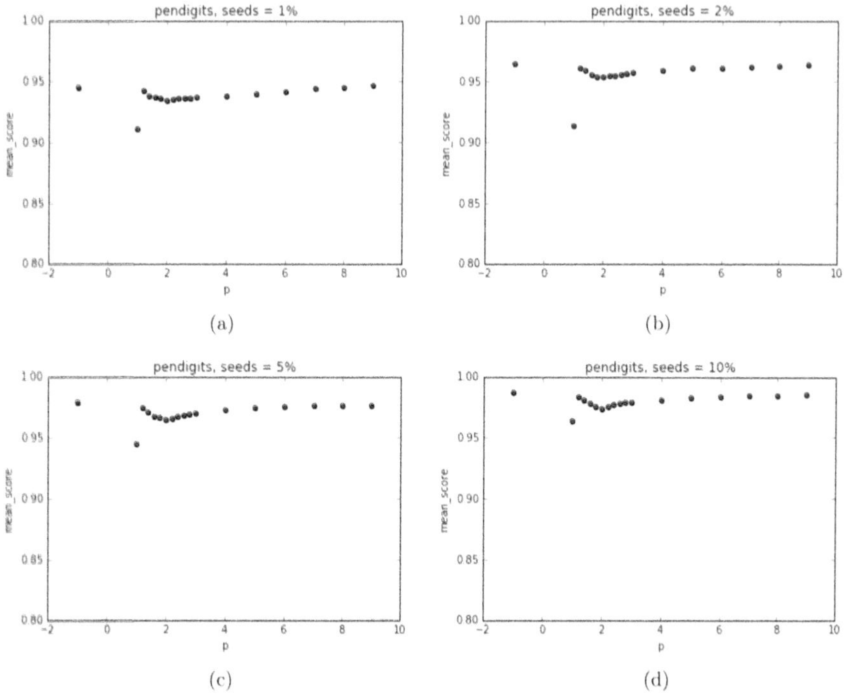

Fig. 5 Overall Caption for the Pendigits Figures

The SiPakMed [10] medical database stands out as a valuable compilation of 4049 annotated cell images from smears. This unique resource encompasses a diversity of cells, providing a detailed representation of cellular variations, from normal states to pathological anomalies. Within its five classes, including "Superficial-Intermediate," "Parabasal," "Koilocytotic," "Dyskeratotic," and "Metaplastic," SiPakMed offers a wealth of data conducive to research and classification.

In Fig. 6, representative examples of cells from the Superficial-Intermediate (a) and Koilocytotic (b) classes in the SiPakMed database are presented.

In this study, our focus is on the binary classification of SiPakMed, specifically on the classes "Superficial-Intermediate," representing normal cells, and "Koilocytotic," characterizing pathological cells. The goal is to explore an innovative method of semi-supervised classification based on partial differential equations. This approach provides a unique perspective by leveraging data characteristics for classification, while introducing parametric flexibility with different values of "p" and "seeds." The results obtained for each combination of parameters provide crucial insights into the performance of our method within the specific context of SiPakMed.

We will examine in detail the significant results obtained for each 'p' and 'seeds,' highlighting key observations and implications for our understanding of cellular structures and pathological anomalies within SiPakMed. These results are conve-

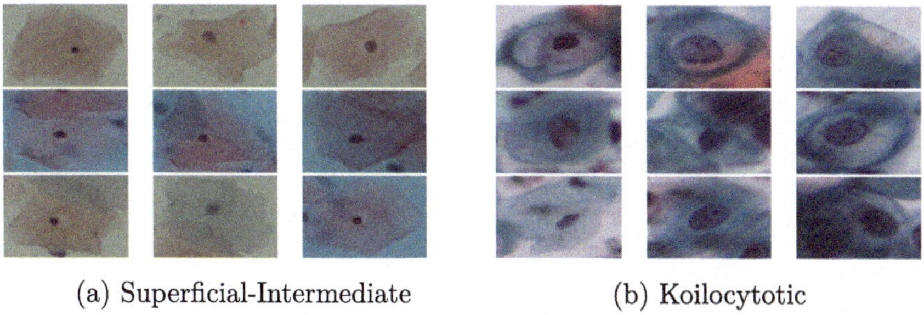

Fig. 6 Representative examples of cells from the Superficial-Intermediate (**a**) and Koilocytotic (**b**) classes in the SiPakMed database

Table 4 Results table for semi-supervised classification: impact of different 'seeds' and 'p' configurations on SiPakMed

% of seeds	$p = 1$	$p = 2$	$p = \infty$
10%	77.5946	76.1905	**79.9756**
5%	77.6557	76.4957	**79.4261**
1%	**76.9231**	75.3968	76.7399

niently summarized in the Table 4, providing insights into the performance of our innovative approach and opening promising avenues for precise cell classification in the medical field.

5 Conclusion

In this study, the Game p-Laplacian on graphs is introduced as an adaptable solution for complex problems in image processing and 3D point cloud data. The extension to semi-supervised classification demonstrates robustness across literature databases and a medical dataset, suggesting potent applications in crucial medical contexts. Future implementation on real medical data aims to enhance validity and clinical relevance. In summary, this research reveals promising applications of the Game p-Laplacian, inspiring ongoing innovative methods for complex problem-solving across diverse domains.

Acknowledgements This work was supported by the Normandy Region and by the European Union (FEDER) through the LoRRI project.

References

1. Barrett, J.W., Liu, W.B.: Finite element approximation of the p-Laplacian. Math. Comput. **61**(204), 523–537 (1993)
2. Carlini, E., Tozza, S.: A scheme for the game p-Laplacian and its application to image inpainting. Appl. Math. Comput. **461**, 128299 (2024)
3. Elmoataz, A., Desquesnes, X., Toutain, M.: On the game p-Laplacian on weighted graphs with applications in image processing and data clustering. Eur. J. Appl. Math. **28**(6), 922–948 (2017)
4. Falcone, M., Vita, S.F., Giorgi, T., Smits, R.G.: A semi-Lagrangian scheme for the game p-Laplacian via p-averaging. Appl. Numer. Math. **73**, 63–80 (2013)
5. Laghrib, A., Afraites, L., Nachaoui, M.: New Trends of Mathematical Inverse Problems and Applications: ICNTAM 2022, Béni Mellal, Morocco, May 19–21, vol. 428. Springer Nature (2023)
6. Lozes, F., Elmoataz, A., Lézoray, O.: PDE-based graph signal processing for 3-d color point clouds: opportunities for cultural heritage. IEEE Signal Process. Mag. **32**(4), 103–111 (2015)
7. Oberman, A.M.: Finite difference methods for the infinity Laplace and p-Laplace equations. J. Comput. Appl. Math. **254**, 65–80 (2013)
8. Parviainen, M.: Notes on tug-of-war games and the p-Laplace equation (2022). arXiv:2208.09732
9. Peres, Y., Schramm, O., Sheffield, S., Wilson, D.: Tug-of-war and the infinity Laplacian. J. Am. Math. Soc. **22**(1), 167–210 (2009)
10. Plissiti, M.E., Dimitrakopoulos, P., Sfikas, G., Nikou, C., Krikoni, O., Charchanti., A.: Sipakmed: a new dataset for feature and image based classification of normal and pathological cervical cells in pap smear images. In: 2018 25th IEEE international conference on image processing (ICIP), pp. 3144–3148. IEEE (2018)
11. Rudd, M.: Statistical exponential formulas for homogeneous diffusion (2014). arXiv:1403.1853

Hybrid Model for Image Denoising Using Variable Exponents: A Novel High-Order PDE Framework with Local-Nonlocal Coupling

Amin Boukdir, Amine Laghrib, and Mourad Nachaoui

Abstract In this paper, we propose a novel hybrid model combining local and non-local methods for effective image denoising. The model utilizes variable exponents $p(x)$ to achieve adaptive diffusion behavior and preserve image features. Nevertheless, the coupling structure of our proposed process, along with the spatial dependence of $p(x)$, poses challenges in theoretical analysis. To address this, we investigate the existence and uniqueness of the solution using the Faedo-Galerkin approximation, a priori estimates, and compactness considerations. Through these theoretical investigations, we demonstrate the existence of a weak solution for the coupled parabolic system. Overall, our model outperforms existing methods in noise reduction and image quality preservation.

Keywords Local enhancement · Nonlocal PDE · Non-linear diffusion · Galerkin's approximation · Reaction-diffusion equations

1 Introduction

Image processing has become essential in modern science and technology, making a significant impact on various fields, including geology, astronomy, computer vision, microscopy, medical imaging, and personal recognition. The quality of images is crucial for many applications that rely on visual data. However, the processes of capturing, recording, transmitting, and acquiring images can introduce various forms of corruption and degradation, such as noise, blurring, and interference. A key task in image processing is denoising, which aims to accurately restore the original images from those affected by noise [1–5]. This process seeks to recover a clean image while preserving important features such as edges, textures, and fine details [6–9]. Mathematically, an observed noisy image u can be modeled as a combination of the true image u_0 and a noise term ε, as follows:

A. Boukdir (✉) · A. Laghrib · M. Nachaoui
EMI, FST, Université Sultan Moulay Slimane, Beni-Mellal, Morocco
e-mail: aminboukdir111@gmail.com

A. Laghrib et al. (eds.), *Computational Methods for Inverse Problems and Applications*, Springer Proceedings in Mathematics & Statistics 498,
https://doi.org/10.1007/978-3-031-89498-5_3

$$u(i) = u_0(i) + \varepsilon(i), \tag{1}$$

where $i \in \mathcal{I}$ represents the pixel index, $u(i)$ is the observed image, $u_0(i)$ is the original image, and $\varepsilon(i)$ represents the noise perturbation. In many cases, $\varepsilon(i)$ is assumed to be independent and identically distributed (i.i.d.) white noise, which complicates the recovery of the original image u_0. Over the years, numerous image denoising techniques have been developed, among which partial differential equation (PDE)-based methods have attracted significant attention due to their solid mathematical foundation and their ability to denoise images without requiring prior training [10]. PDE-based methods are typically classified as local methods because they rely on local image gradients to drive the diffusion process. One of the earliest models used for image denoising is based on the heat equation, which describes isotropic diffusion:

$$\frac{\partial u}{\partial t} = c\Delta u, \quad u(0) = u_0, \tag{2}$$

where u is the image at time t, Δu is the Laplace operator, and $c > 0$ is a constant controlling the diffusion speed. The solution to this equation is equivalent to applying a Gaussian filter to the image, leading to smoothing. However, this model introduces excessive blurring, as it tends to smooth both the noise and essential structures, such as edges [7]. To address the blurring issue, Perona and Malik proposed a nonlinear anisotropic diffusion model:

$$\frac{\partial u}{\partial t} = \operatorname{div}(g(|Du|^2)Du). \tag{3}$$

The Perona-Malik model allows for diffusion to be stronger in smooth regions (where $|Du|$ is small) and weaker near edges (where $|Du|$ is large), preserves discontinuities, but performs poorly on flat and textured regions. This model has got many variants and extensions. En particular, the total variation of the image (TV) which is related to the mean curvature motion and to the PM equation when $g(|Du|^2) = \frac{1}{|Du|}$, leads to the minimization of $TV(u) = \int |Du|$, subject to some constraints [10]. The gradient descent scheme of this energy reads, at least formally,

$$\frac{\partial u}{\partial t} = \operatorname{div}\left(\frac{Du}{|Du|}\right). \tag{4}$$

The main advantage of the TV model for image denoising is the powerful ability to restore edges and homogeneous regions. Many extension of TV-based denoising algorithms has been developed [11–13]. However, these approaches lead to the staircase effect and it fails to preserve textured structures and repeated patterns in images. An effective solution to this drawback is obtained from increasing the order of the derivative. From this point of view, various fourth-order PDEs appeared [14–17]. For example, the following fourth-order PDE proposed in [18] for noise removal

$$\frac{\partial u}{\partial t} = \Delta\left(g(\Delta u)\Delta u\right), \tag{5}$$

where g denotes a nonnegative monotonically decreasing function. However, the fourth-order PDEs suffer from the blurring effect near sharp edges due to the possible over-smoothing. To tackle the staircase effect while avoiding the creation of blur and maintain the structure-preserving ability, many authors have presented fractional-order models [19–21] inspired from the model introduced by Bai and Feng in [22].

$$\frac{\partial u}{\partial t} = -\left(D_x^\alpha\right)^*\left(c\left(|D^\alpha u|^2\right)D_x^\alpha u\right) - \left(D_y^\alpha\right)^*\left(c\left(|D^\alpha u|^2\right)D_y^\alpha u\right). \tag{6}$$

Despite the existence of numerous PDE-based methodologies, it is extremely challenging to create a PDE that completely overcomes the aforementioned drawbacks. As a result, in the past ten years, certain methods have emerged that appear to learn PDEs for particular vision tasks [23–25]. Since there is no human intervention, learning-based PDEs using training data have an advantage over conventional PDEs. The aforementioned approaches, however, rely solely on the local information of the image, which limits their ability to preserve textured structures and recurring patterns. To address this limitation, nonlocal diffusion models were introduced [26]. A typical nonlocal model takes the form:

$$\frac{\partial u}{\partial t} = \int_\Omega \mathcal{J}(x-y)|u(y,t)-u(x,t)|^{p-2}(u(y,t)-u(x,t)), dy, \tag{7}$$

where $\mathcal{J}(x-y)$ is a kernel function that measures the similarity between patches centered at x and y. Nonlocal methods are particularly effective at capturing long-range dependencies and repetitive patterns in images. However, they can introduce residual noise in homogeneous areas. To leverage the strengths of both local and nonlocal methods, recent research has focused on coupled diffusion models that combine local and nonlocal diffusion operators. For example, the model proposed by Shi et al. [27] couples a local p-Laplacian with a nonlocal p-Laplacian:

$$\frac{\partial u}{\partial t} = \operatorname{div}\left((1-v(x))|\nabla u|^{p-2}\nabla u\right) + \alpha\int_\Omega \frac{v(x)+v(y)}{2}\mathcal{J}(x-y)|u(y,t)-u(x,t)|^{p-2}(u(y,t)-u(x,t)), dy, \tag{8}$$

where $v(x)$ is a space-varying function that controls the balance between local and nonlocal diffusion. Motivated by the limitations of existing approaches, we propose a novel reaction-diffusion model that combines local and nonlocal methods. Our model integrates the local $p(x)$-Laplacian operator with a nonlocal $p(x)$-Laplacian, where the variable exponent $p(x)$ adapts to the image structure. This hybrid approach aims to harness the advantages of both local and nonlocal methods while minimizing their drawbacks, such as over-smoothing and loss of texture. The variable exponent $p(x)$ enables better control of diffusion near edges and faster diffusion in homogeneous regions, making the model highly adaptable to the multi-scale nature of images. The rest of the paper is organized as follows. Section 2 describes the proposed model and presents a brief statement on the main results. Section 3 is devoted to the numerical approximation of the proposed reaction-diffusion PDE, while Sect. 4 presents and discusses some numerical experiment results for both benchmark and real images.

2 Description of the Proposed Model

To exploit the advantages of both local and nonlocal image denoising models, we decompose the target image u into two components: the smooth part (denoted by Y) and the texture part (denoted by X). Subsequently, the proposed diffusion process is separately applied to these two image components, as described below:

$$\begin{cases} \frac{\partial X}{\partial t} = \operatorname{div}\left(|\nabla X|^{p(x)-2}\nabla X\right) - \nabla\cdot(\eta(x)\nabla Y) & \text{in } \Omega_T, \\ \frac{\partial Y}{\partial t}(t, x) = \int_\Omega \mathfrak{J}(x-y)|Y(y,t)-Y(x,t)|^{p(x)-2}\left(Y(y,t)-Y(x,t)\right)dy & \\ -(u_0 - \nabla\cdot(\eta(x)\nabla X)) & \text{in } \Omega_T, \\ X(0,x) = u_0, & \text{in } \Omega, \\ Y(0,x) = 0, & \text{in } \Omega, \end{cases} \tag{9}$$

With Neumann boundary conditions: $\langle|\nabla X|^{p(x)-2}\nabla X, n\rangle = \langle\eta(x)\nabla Y, n\rangle = \langle\eta(x)\nabla X, n\rangle = 0$, and $\int_{\partial\Omega_T} Y(y,t) - Y(x,t)|^{p(x)-2}\left(Y(y,t) - Y(x,t)\right) n\, ds\, dt = 0$, where $\Omega_T = \Omega\times(0, T]$, $\Omega\subset\mathbb{R}^2$, is a bounded smooth domain with Lipschitz continuous boundary $\partial\Omega$. The exponent of nonlinearity p is a given function, $p:\Omega\mapsto\left[p^-, p^+\right]\subset(1,2)$, which is defined by:

$$p(x) = 1 + \frac{1}{1+|x|+k_2\left|\nabla G_\sigma * u_0\right|^2}, \quad k_2 > 0, \sigma > 0.$$

The kernel $\mathfrak{J}:\mathbb{R}^2\to\mathbb{R}$ is a nonnegative continuous radial function with compact support, $\mathfrak{L}(0) > 0$, and $\int_{\mathbb{R}^2}\mathfrak{L}(z)dz = 1$. Given a noisy image u_0, the solution $u(x,\ t)$ is the restored image with scale variable t. The parameters of the model satisfy the following hypothesis:

(H_1) : The function u_0 is in $L^2(\Omega)$.
(H_2) : Assume that the function $p(x):\Omega\to(1,+\infty)$ satisfies $1 < p^- \le p(x) \le p^+ < \infty$, where p^-, $p^+$are given fixed constants.
(H_3) : The constant λ is positive.
(H_4) : $\eta(x)$ is a bounded, positive function in $L^\infty(\Omega)$.

3 Existence and Uniqueness

We begin by a weak formulation of our model (9), where the solution is interpreted according to the following definition.

Definition 1 A pair of measurable functions (X, Y), $X : \Omega_T \to \mathbb{R}$ and $Y : \Omega_T \to \mathbb{R}$ is a weak solution of (9) in $\Omega_T \times \Omega_T$, if $X \in \mathbf{W}(\Omega_T) \cap L^\infty\left(0, T; L^2(\Omega)\right)$ and $Y \in \mathbf{S}(\Omega_T) \cap L^\infty\left(0, T; H^1(\Omega)\right)$, such that

$$\begin{cases} \int_{\Omega_T} X\phi_t \, dx \, dt - \int_\Omega X\phi dx|_{\tau=0}^{\tau=t} - \int_{\Omega_T} |\nabla X|^{p(x)-2}\nabla X \cdot \nabla\phi dx \, dt + \int_{\Omega_T} \nabla\phi \cdot (\eta(x)\nabla Y) \, dx \, dt = 0, \\ \int_{\Omega_T} Y\varphi_t \, dx \, dt - \int_\Omega Y\varphi dx|_{\tau=0}^{\tau=t} + \int_{\Omega_T}\int_\Omega \mathfrak{J}(x-y)|DY(x, y, t)|^{p(x)-2} DY(x, y, t) D\varphi dy dx dt \\ + \int_{\Omega_T} u_0\varphi dx \, dt - \int_{\Omega_T} \nabla\varphi \cdot (\eta(x)\nabla X) \, dx \, dt = 0, \\ \forall t \in [0, T], \phi \in \mathbf{W}(\Omega_T) \cap L^\infty\left(0, T; L^2(\Omega)\right), \\ \phi_t \in \mathbf{W}'(\Omega_T), \forall\varphi \in \mathbf{S}(\Omega_T) \cap L^\infty\left(0, T; H^1(\Omega)\right), \varphi_t \in \mathbf{S}'(\Omega_T). \end{cases} \tag{10}$$

the following lemmas hold and their proof is similar to the ones mentioned in [28]:

Lemma 1 *For every $T < \infty$ and $m \in \mathbb{N}$ the system (9) has a solution $\left\{c_k^{(m)}(t), d_k^{(m)}(t)\right\}_{k=1}^m$ on the interval $(0, T)$ and the corresponding functions $X^{(m)}$ and $Y^{(m)}$ satisfy the estimate*

$$\begin{aligned} &\left\|X^{(m)}\right\|^2_{L^\infty(0,T;L^2(\Omega))} + \left\|V^{(m)}\right\|^2_{L^\infty(0,T;L^2(\Omega))} \\ &+ \int_0^t \int_\Omega \delta_0 \left|\nabla X^{(m)}\right|^{p(x)} dx \, d\tau + \lambda \int_0^t \int_\Omega \left|\nabla Y^{(m)}\right|^2 dx \, d\tau \le C\|u_0\|^2_{L^2(\Omega)}, \end{aligned} \tag{11}$$

where the constant C depends only on T, λ and

$$\delta_0 = M\left(G_\sigma, \|u_0\|_{L^2(\Omega)}\right) \le g(x) \le 1,$$

where $M > 0$ is a constant.

Lemma 2 *Under the assumptions of Lemma 1, for every $m \in \mathbb{N}$ we have $X_t^{(m)} \in \mathbf{W}'(\Omega_T)$ and $Y_t^{(m)} \in \mathbf{S}'(\Omega_T)$ such that*

$$\left\|U_t^{(m)}\right\|_{\mathbf{W}'(\Omega_T)} \le C\|u_0\|_{L^2(\Omega)},$$
$$\left\|V_t^{(m)}\right\|_{\mathbf{Y}'(\Omega_T)} \le C\|u_0\|_{L^2(\Omega)}.$$

We have then the following Theorem, that ensures the existence and uniqueness of the solution

Theorem 1 *Let $\Omega \subset \mathbb{R}^N$ be a bounded domain with smooth boundary. If $u_0 \in L^2(\Omega)$, then the problem (9) admits a unique weak solution (X, Y) such that*

$$\|X(\cdot,t)\|^2_{L^\infty(0,T;L^2(\Omega))} + \|Y(\cdot,t)\|^2_{L^\infty(0,T;L^2(\Omega))} + \int_0^t \int_\Omega |\nabla X|^{p(x)} \mathrm{d}x \, \mathrm{d}\tau + \lambda \|\nabla Y(\cdot,t)\|^2_{L^\infty(0,T;L^2(\Omega))} \leq C\|u_0\|^2_{L^2(\Omega)},$$

where the constant C is dependent of T, η, G_σ and $\|u_0\|_{L^2(\Omega)}$. Moreover, $X_t \in \mathbf{W}'(\Omega_T)$ and $Y_t \in \mathbf{S}'(\Omega_T)$.

Remark 1 The proof of the theorem 1 is similar to the proof given in [28].

4 Discrete Setting

This section deals with the numerical approximation of the proposed model. In fact, to compute numerically the problem (10), we carry out a fully discretization of the model using finite element method.

We start by temporary time descretization defined as follow:

$$[0, T] \simeq \{t_n\} = n\Delta t, \quad \Delta t = \frac{T}{N} \text{ and } X^n = X(t_n, x), \ Y^n = Y(t_n, x) \ n \in \{0, \ldots, N\}, \quad x \in \Omega$$

Denote X^n and Y^n the approximations of the two components of $X(t_n, x)$, then we define the discrete approximations:

$$\frac{\partial X(t,x)}{\partial t} \simeq \frac{X^{n+1} - X^n}{\Delta t}, \quad \frac{\partial Y(t,x)}{\partial t}, \simeq \frac{Y^{n+1} - Y^n}{\Delta t}. \tag{12}$$

$$X^0 = X(0, x) = u_0(x), \quad Y^0 = Y(0, x) = v_0(x) \text{ Initial condition.}$$

As the problem is coupled system with nonlinear lolcal and nonlocal terms, we opt for the implicit Euler schema. Consequently, for $n \in \{1, \ldots, N\}$ find $(X^n, Y^n) \in U(\Omega) \times S(\Omega)$ solution of

$$\int_\Omega \frac{X^n - X^{n-1}}{\Delta t} \phi \, dx + \int_\Omega |\nabla X^n|^{p(x)-2} \nabla X^n \cdot \nabla \phi \, dx = \int_\Omega \eta(x) \nabla Y^{n-1} \cdot \nabla \phi \, dx.$$

$$\int_\Omega \frac{Y^n - Y^{n-1}}{\Delta t} \varphi \, dx + \int_\Omega \int_\Omega \mathfrak{J}(x-y) |Y^{n-1}(y) - Y^{n-1}(x)|^{p(x)-2} (Y^n(y) - Y^n(x)) \, dy \, dx = \int_\Omega (u_0 - \nabla \cdot (\eta(x) \nabla X^{n-1})) \varphi \, dx.$$

$$\int_{\Omega} X^n \phi \, dx + \Delta t \int_{\Omega} |\nabla X^{n-1}|^{p(x)-2} \nabla X^n \cdot \nabla \phi \, dx = \int_{\Omega} X^{n-1} \phi + \Delta t \int_{\Omega} \eta(x) \nabla Y_h^{n-1} \cdot \nabla \phi \, dx, \tag{13}$$

$$\begin{aligned} \int_{\Omega} Y^n \varphi \, dx - \Delta t \int_{\Omega} \int_{\Omega} \mathfrak{J}(x-y) |DY^{n-1}|^{p(x)-2} \left(DY^n\right) D\varphi \, dy \, dx = \int_{\Omega} Y^{n-1} \varphi \, dx + \Delta t \int_{\Omega} u_0 \varphi \, dx \\ + \Delta t \int_{\Omega} \eta(x) \nabla X^{n-1} \cdot \nabla \varphi \, dx. \end{aligned} \tag{14}$$

It remains now to discretize the local and nonlocal Eqs. (13) and (14). To do so, we consider the method represented by D'Elia et al. [29, Chap. 3].

We consider finite element approximations in order to define appropriate finite-dimensional energy spaces Z^h, where $h > 0$ is a parameter determining the grid size of the underlying finite element mesh.

For simplicity we assume that Ω is a polyhedral domain and that the interaction sets $\{S(\mathbf{x})\}_{\mathbf{x} \in \mathbb{R}^d}$ are simply connected. The first requirement implies that we can exactly triangulate Ω into finite (polyhedral) elements $\mathcal{T}^h = \{\mathcal{E}_k\}_{k=1}^K$ with

$$\bar{\Omega} = \bigcup_{k=1}^{J} \overline{\mathcal{E}}_k.$$

We assume that the corners of the boundary $\partial\Omega$ are vertices of the finite element mesh.

Let $\left\{\mathbf{x}_j\right\}_{j=1}^J$ denote the mesh vertices of the triangulation $\mathcal{T}^h$.

Now let us assign to each vertex of the triangulation $\mathcal{T}^h$ a continuous piecewise polynomial function $\Theta_j : \Omega \to \mathbb{R}$ leading to a family of basis function $\left\{\Theta_j\right\}_{j=1}^J$. We require $\Theta_j\left(\mathbf{x}_j\right) = \delta_{jj'}$, where $\delta_{jj'}$ denotes the Kroenecker delta function, and define the finite-dimensional finite element spaces for $j, j' = 1, \ldots, J_{\Omega}$.

$$Z^h := \operatorname{span}\left\{\Theta_j : 1 \le j \le J\right\}$$

The discretization of the system (13) and (14) using the finite element, leads to the approximate problem: Find $U_h^n \in Z^h$

$$a_1\left(X_h^n, \Theta_j\right) = \int_{\Omega} X_h^n \Theta_j \, \mathrm{d}x + \Delta t \int_{\Omega} |\nabla X_h^{n-1}|^{p(x)-2} \nabla U_h^n \cdot \nabla \Theta_j \mathrm{d}x \tag{15}$$

$$l_1(\Theta_j) = \int_{\Omega} X_h^{n-1} \Theta_j + \Delta t \int_{\Omega} \eta(x) \nabla Y_h^{n-1} \cdot \nabla \Theta_j, \tag{16}$$

where $a_1(\cdot, \cdot)$ and $l_1(\cdot)$ denotes respectively the bilinear and linear form correspond to the variational Eq. (13). The finite element approximation $X_h^n(\mathrm{x})$ has the form

$$X_h^n(\mathbf{x}) = \sum_{j=1}^{J} X_j^n \Theta_j(\mathbf{x}), \quad \text{and } Y_h^n(\mathbf{x}) = \sum_{j=1}^{J} Y_j^n \Theta_j(\mathbf{x}).$$

In the same way we define the bilinear and linear form correspond to the variational Eq. (14), find $Y^n \in Z^h$

$$a_2\left(Y_h^n, \Theta_j\right) = \int_\Omega Y_h^n \Theta_j - \Delta t \int_\Omega \int_\Omega \mathfrak{J}(x-y)|DY_h^{n-1}|^{p(x)-2}\left(DY_h^n\right) D\Theta_j \quad (17)$$

$$l_2(\Theta_j) = \int_\Omega Y_h^{n-1}\Theta_j + \Delta t \int_\Omega u_0 \Theta_j + \Delta t \int_\Omega \eta(x) \nabla X_h^{n-1} \cdot \nabla \Theta_j. \quad (18)$$

Solving the system (13) and (14) leads to

$$\text{for } n = 1, \ldots, N \text{ solve the linear system}$$
$$A_1 U_h^n = b_1 \quad (19)$$
$$A_2 V_h^n = b_2, \quad (20)$$

where

$$A_1 := a_1(\Theta_i, \Theta_j) = \int_\Omega \Theta_i \Theta_j \, \mathrm{d}x + \Delta t \int_\Omega |\nabla X_h^{n-1}|^{p(x)-2} \nabla \Theta_i \cdot \nabla \Theta_j \mathrm{d}x, \quad (21)$$

$$b_1 := l_1(\Theta_j) = \int_\Omega X_h^{n-1}\Theta_j + \Delta t \int_\Omega \eta(x) \nabla Y_h^{n-1} \cdot \nabla \Theta_j \quad (22)$$

$$A_2 := a_2(\Theta_i, \Theta_j) = \int_\Omega \Theta_i \Theta_j - \Delta t \int_\Omega \int_\Omega \mathfrak{J}(x-y)|DY^{n-1}|^{p(x)-2}(D\Theta_i) D\Theta_j \quad (23)$$

$$b_2 := l_2(\Theta_j) = \int_\Omega Y_h^{n-1}\Theta_j + \Delta t \int_\Omega u_0 \Theta_j + \Delta t \int_\Omega \eta(x) \nabla X_h^{n-1} \cdot \nabla \Theta_j. \quad (24)$$

To solve the system (20) we opt for the LGMRES algorithm, which is designed to prevent the convergence slowdown in restarted GMRES. It is common for it to outperform GMRES(m) in some way for comparable memory requirements, or at least not much worse [30].

In the following section, we give some numerical experiments to confirm the efficiency of the proposed PDE.

5 Numerical Results

In this section, we explore the experimental results of the proposed PDE model, comparing its performance to established denoising techniques. Our iterative algorithm has demonstrated successful convergence, which we validate using a specific criterion. The stopping condition for our iterative process is defined as follows:

$$\frac{\|X^{k+1}(T) - X^k(T)\|_{L^2(\Omega)}}{\|X^k(T)\|_{L^2(\Omega)}} < 10^{-6}.$$

For our evaluations, we utilize the Berkeley Dataset, a comprehensive collection of diverse natural images. We commence our analysis with two tests designed to highlight the recovery of the component u through our method. The parameters guiding our algorithm are set as follows: $\lambda = 0.2$, $\alpha = 0.6$, $\mu = 0.15$, $\xi = 1.5$, $\zeta = 3.2$, $\epsilon = 0.0004$, $\delta = 1/2$, and we execute $N = 1000$ iterations for the tests.

We conduct an analysis using four test images, each exhibiting unique features and subjected to varying levels of Gaussian noise, specifically with parameters $r = 0.02$, $r = 0.03$, $r = 0.04$, and $r = 0.05$. Our evaluation centers on the proposed model's effectiveness in preserving intricate details and textures while efficiently mitigating noise. In particular, we investigate the model's performance in retaining sharp edges and structural patterns, especially in areas with significant anisotropic characteristics.

For comparison, we assess our model against the fractional-order model (FOM) [21] and the fourth-order PDE (FOPDE) proposed in [16]. Figure 1 showcases the denoised results obtained from our model alongside those from other denoising methods. The findings across all tests demonstrate that our model excels at maintaining edges and textures while achieving substantial noise reduction, even when applied to a diverse range of images-whether synthetic, geometric, or textured. In Fig. 2, we present the evolution of the logarithmic relative error over the course of the iterative process. This graph illustrates how the error decreases as the algorithm progresses compared to other algorithms for the two tests: *Mosq* and *Man*, indicating improved convergence towards the true solution. The diminishing error values reflect the effectiveness of our method in refining the image reconstruction and highlight the stability of the proposed approach throughout the iterations. In Table 1, we present a quantitative comparison of four examples using two key metrics: the Peak Signal-to-Noise Ratio (PSNR), which measures the ratio of signal power to noise power in the image, and the Structural Similarity Index (SSIM), which assesses the perceptual similarity between two images. The results indicate that our proposed method consistently outperformed the competing techniques across all test cases. We conclude with a final real-world test of the *Fly logo* image. This test serves to demonstrate the practical application of our method in a real imaging scenario, showcasing its effectiveness in preserving image quality while reducing noise. The results in Fig. 3 highlight the robustness of our approach, confirming its suitability for various types of images.

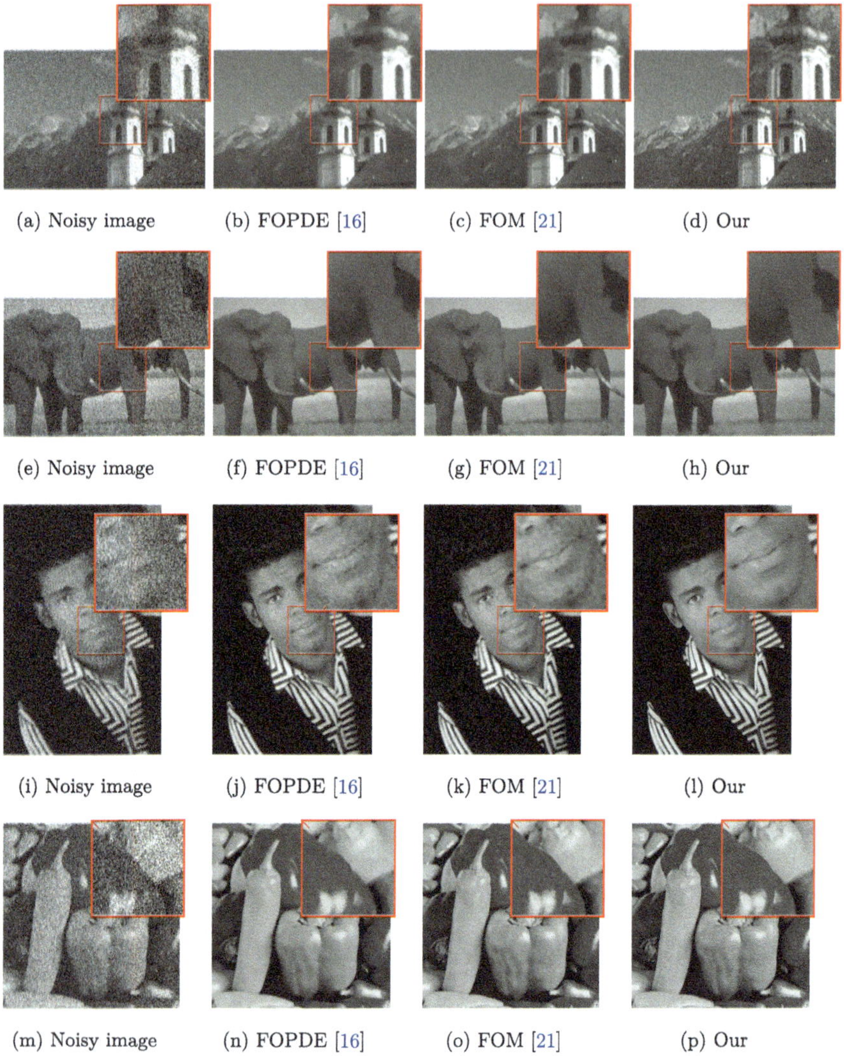

Fig. 1 Reconstruction of natural tests: *Mosq*, *Elephant*, *Man* and *Peppers*

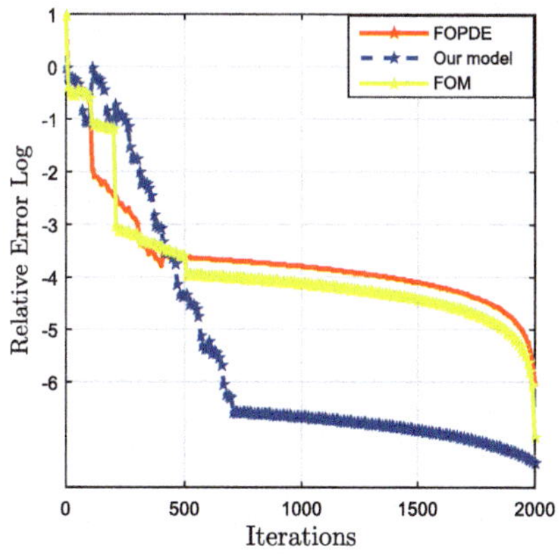

(a) *Mosq*

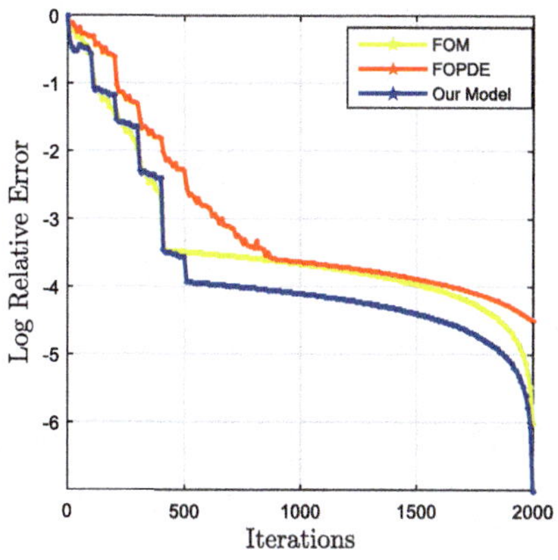

(b) *Man*

Fig. 2 The variation of the Log Relative Error with respect to the iteration number for the two previous selected images

Table 1 The PSNR and SSIM values for the evaluated methods on four additional test images

Image	Method	PSNR	SSIM	Image	Method	PSNR	SSIM
Mosq	FOPDE [16]	31.45	0.910	*Elephant*	FOPDE [16]	29.12	0.815
	FOM [21]	32.00	0.915		FOM [21]	29.00	0.820
	Our	33.50	0.940		Our	30.25	0.845
Man	FOPDE [16]	28.75	0.750	*Peppers*	FOPDE [16]	28.99	0.800
	FOM [21]	29.50	0.770		FOM [21]	29.45	0.810
	Our	30.10	0.820		Our	29.75	0.830

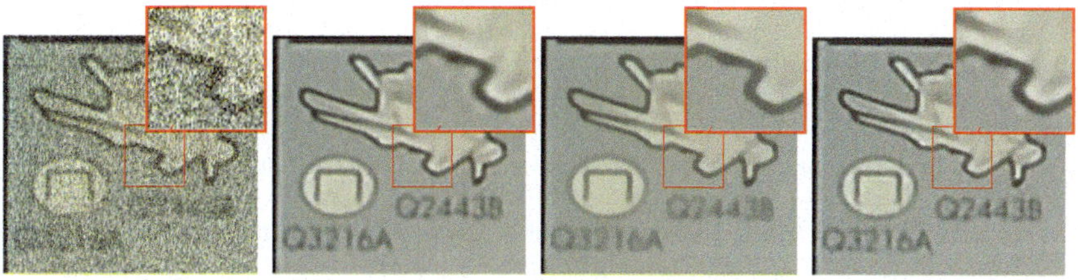

(a) Noisy image (b) FOPDE [16] (c) FOM [21] (d) Our

Fig. 3 The variation of the Log Relative Error with respect to the iteration number for the two previous selected images

References

1. Zaabouli, Z., El Hakoume, A., Afraites, L., Laghrib, A.: A novel coupled p (x) and fractional PDE denoising model with theoretical results. Int. J. Comput. Math. **101**(11), 1300–1325 (2024)
2. Boukdir, A., Laghrib, A., Nachaoui, M.: Time-fractionated anisotropic osmosis with bilateral total variation for effective shadow mitigation in image processing. Appl. Math. Model. **137**, 115671 (2025)
3. Laghrib, A., Afraites, L., Nachaoui, M.: New trends of mathematical inverse problems and applications
4. Laghrib, A., Limami, F.-E., Nachaoui, A.: Enhanced osmosis model with bilateral total variation for effective shadow removal. Comput. & Math. Appl. **174**, 71–83 (2024)
5. El Hakoume, A., Laghrib, A., Hadri, A., Afraites, L.: An optimal fluid optical flow registration for super-resolution with lamé parameters learning. J. Optim. Theory Appl. **197**(2), 508–538 (2023)
6. Aubert, G., Kornprobst, P., Aubert, G.: Mathematical Problems in Image Processing: Partial Differential Equations and the Calculus of Variations, vol. 147. Springer (2006)
7. Perona, P., Malik, J.: Scale-space and edge detection using anisotropic diffusion. IEEE Trans. Pattern Anal. Mach. Intell. **12**(7), 629–639 (1990)
8. Nachaoui, A., Laghrib, A., Hakim, M.: Mathematical Control and Numerical Applications, vol. 372. Springer (2021)
9. Alahyane, M., Hakim, A., Laghrib, A., Raghay, S.: An improved SIMPLEC scheme for fluid registration. Math. Model. Anal. **28**(1), 71–90 (2023)
10. Rudin, L.I., Osher, S., Fatemi, E.: Nonlinear total variation based noise removal algorithms. Phys. D Nonlinear Phenom. **60**(1–4), 259–268 (1992)
11. Laghrib, A.: A comparative study between tv, tv2, btv and combined models for the multi-frame super-resolution. Adv. Math. Model. & Appl. **5**(1), 80–94 (2020)
12. Nachaoui, M.: Parameter learning for combined first and second order total variation for image reconstruction. Adv. Math. Model. & Appl. **5**(1), (2020)
13. Nachaoui, M., Afraites, L., Laghrib, A.: A regularization by denoising super-resolution method based on genetic algorithms. Signal Process. Image Commun. **99**, 116505 (2021)
14. Afraites, L., Hadri, A., Laghrib, A., Nachaoui, M.: A high order PDE-constrained optimization for the image denoising problem. Inverse Probl. Sci. Eng. **29**(12), 1821–1863 (2021)
15. Hadri, A., Afraites, L., Laghrib, A., Nachaoui, M.: A novel image denoising approach based on a non-convex constrained PDE: application to ultrasound images. Signal Image Video Process. **15**, 1057–1064 (2021)
16. Hadri, A., Khalfi, H., Laghrib, A., Nachaoui, M.: An improved spatially controlled reaction-diffusion equation with a non-linear second order operator for image super-resolution. Nonlinear Anal. R. World Appl. **62**, 103352 (2021)
17. Laghrib, A., Afraites, L., Hadri, A., Nachaoui, M.: A non-convex PDE-constrained denoising model for impulse and gaussian noise mixture reduction. Inverse Probl. Imaging **17**(1), 23–67 (2023)
18. You, Y.-L., Kaveh, M.: Fourth-order partial differential equations for noise removal. IEEE Trans. Image Process. **9**(10), 1723–1730 (2000)
19. Afraites, L., Hadri, A., Laghrib, A.: A denoising model adapted for impulse and gaussian noises using a constrained-PDE. Inverse Probl. **36**(2), 025006 (2020)
20. Janev, M., Pilipovic, S., Atanackovic, T., Obradovic, R., Ralevic, N.: Fully fractional anisotropic diffusion for image denoising. Math. Comput. Model. **54**(1–2), 729–741 (2011)
21. Nachaoui, M., Laghrib, A.: An improved bilevel optimization approach for image super-resolution based on a fractional diffusion tensor. J. Frankl. Inst. **359**(13), 7165–7195 (2022)
22. Bai, J., Feng, X.-C.: Fractional-order anisotropic diffusion for image denoising. IEEE Trans. Image Process. **16**(10), 2492–2502 (2007)
23. Afraites, L., Hadri, A., Laghrib, A., Nachaoui, M.: A weighted parameter identification PDE-constrained optimization for inverse image denoising problem. Vis. Comput. **38**(8), 2883–2898 (2022)

24. Ghazdali, A., Hadri, A., Laghrib, A., Nachaoui, M.: A blind poisson–gaussian noise separation using learning copula densities. Circuits Syst. Signal Process. 1–27 (2023)
25. Nachaoui, M., Afraites, L., Hadri, A., Laghrib, A.: A non-convex non-smooth bi-level parameter learning for impulse and gaussian noise mixture removing. Commun. Pure Appl. Anal. **21**(4), 1249–1291 (2022)
26. Andreu, F., Mazon, J.M., Rossi, J.D., Toledo, J.: A nonlocal p-laplacian evolution equation with neumann boundary conditions. Journal de mathematiques pures et appliquees **90**(2), 201–227 (2008)
27. Shi, K.: Coupling local and nonlocal diffusion equations for image denoising. Nonlinear Anal. R. World Appl. **62**, 103362 (2021)
28. Boukdir, A., Nachaoui, M., Laghrib, A.: Hybrid variable exponent model for image denoising: a nonstandard high-order PDE approach with local and nonlocal coupling. J. Math. Anal. Appl. **536**(2), 128245 (2024)
29. D'Elia, M., Gunzburger, M., Vollmann, C.: A cookbook for approximating Euclidean balls and for quadrature rules in finite element methods for nonlocal problems. Math. Model. Methods Appl. Sci. **31**(08), 1505–1567 (2021)
30. Baker, A.H., Jessup, E.R., Manteuffel, T.: A technique for accelerating the convergence of restarted GMRES. SIAM J. Matrix Anal. Appl. **26**(4), 962–984 (2005)

A Semi-analytical Solution for Inverse Cauchy Problems

Abdeljalil Nachaoui, Sudad M. Rasheed, and Gulnar W. Sadiq

Abstract This paper investigates the numerical solution of a Cauchy problem for harmonic functions modeled by the Fredholm integral equation, In the context of solving this problem, we explore the formulation of the Fredholm equation and apply a regularized Fourier-based method to approximate the solution. The study first introduces the problem setup and the integral equation formulation, then describes the numerical methods employed to solve it. We analyze the influence of key parameters such as the number of Fourier terms N, the number of quadrature points M, and the regularization parameter α on the accuracy and convergence of the solution. The findings provide practical insights for solving Fredholm equations using numerical methods and highlight the importance of parameter tuning for achieving accurate and stable solutions.

Keywords Inverse Cauchy problem · Semi-analytical approach · Fredholm integral equation · Legendre polynomial expansion

1 Introduction

Electrostatic imaging, a key technique in nondestructive testing of metallic disks, involves solving an inverse boundary value problem tied to the Laplace equation in two dimensions. This method aims to determine the unknown shape of an inclusion within a conductive material by imposing additional Cauchy data, such as voltage

A. Nachaoui (✉)
Laboratoire de Mathématiques Jean Leray, Nantes Université, Nantes, France
e-mail: Abdeljalil.Nachaoui@univ-nantes.fr

S. M. Rasheed · G. W. Sadiq
Mathematical Sciences Department, College of Basic Education, University of Sulaimani, Sulaymaniyah, Iraq
e-mail: sudad.rasheed@univsul.edu.iq

G. W. Sadiq
e-mail: gulnar.sadiq@univsul.edu.iq

A. Laghrib et al. (eds.), *Computational Methods for Inverse Problems and Applications*, Springer Proceedings in Mathematics & Statistics 498,
https://doi.org/10.1007/978-3-031-89498-5_4

and current, on the accessible external boundary. This scenario gives rise to an inverse Cauchy problem, which depends entirely on boundary data [1–14]. However, it is well-established that such problems are inherently ill-posed.

Mathematically, let $u(x, y)$ represent the potential function, with $u_n(x, y)$ as its normal derivative at any point $(x, y) \in \mathbb{R}^2$. The normal derivative $u_n(x, y)$ is evaluated on the outer boundary Γ_1, a circle of radius r_1 within an annular domain Ω. The objective is to solve the two-dimensional Laplace equation $\Delta u(x, y) = 0$, determining $u(x, y)$ on an internal circle Γ_2 of radius $r_2 < r_1$. This configuration is particularly significant for inverse electrostatic imaging applications, such as human electroencephalography [15]. The task involves resolving a disk-based equation using overspecified Cauchy boundary data.

Approaches to solving inverse Cauchy problems generally fall into two main categories: iterative and direct methods [14, 16–29]. Iterative techniques, such as the conjugate gradient method [30], begin with an initial estimate and refine the solution iteratively through minimization or well-posed sub-problems. While iterative methods are computationally demanding, recent advancements have focused on reducing iteration counts [31, 32]. However, achieving convergence in a single iteration remains unattainable.

Direct methods, in contrast, are computationally efficient, typically requiring only one solver. Examples include the quasi-reversibility method [33–35], which reduces computation time but may be prone to numerical instability.

Both methods require discretizing the domain where problem is studied, often through finite difference (FDM), finite element (FEM), or boundary element methods (BEM). Challenges arise from domain meshing requirements, prompting interest in meshless methods. These approaches, which substitute traditional meshing with arbitrarily distributed nodes, offer an alternative framework [13, 26, 36–44].

Among meshless techniques, polynomial expansion methods are particularly notable and have become very familiar and increasingly used [37, 40, 43, 45–49]. These methods provide flexible and robust algorithms suitable for most boundary value problems (BVPs) and have been successfully integrated with multi-scale preconditioning to enhance performance [43, 48–50].

1.1 The Cauchy Problem

Consider an annular domain bounded by an inner radius r_2 and an outer radius r_1, where $r_2 < r_1$. Let $u_x = \frac{\partial u}{\partial x}$ and $u_{xx} = \frac{\partial^2 u}{\partial x^2}$, with x representing either r or θ. The Cauchy problem in this domain is formulated as follows:

$$u_{rr}(r,\theta) + \frac{1}{r}u_r(r,\theta) + \frac{1}{r^2}u_{\theta\theta}(r,\theta) = 0, \quad r \in (r_2, r_1),\ \theta \in (0, 2\pi), \tag{1}$$

$$u(r_1,\theta) = g(\theta), \quad \theta \in [0, 2\pi], \tag{2}$$

$$u_r(r_1,\theta) = h_1(\theta), \quad \theta \in [0, 2\pi]. \tag{3}$$

2 The Fredholm Equation Formulation

To derive a Fredholm integral equation of the first kind for the unknown function $f(\theta)$ on the inner boundary, we replace the boundary condition in Eq. (3) with the condition:

$$u(r_2, \theta) = f(\theta), \quad \theta \in [0, 2\pi], \tag{4}$$

where $f(\theta)$ is the function to be determined. Assuming that $f(\theta)$ in Eq. (4) is unknown and $h_1(\theta)$ in Eq. (3) is overspecified, the goal is to determine $f(\theta)$.

Using separation of variables with Eqs. (1), (2), and (4), the solution takes the form:

$$\begin{aligned} u(r,\theta) = A_0 \ln\left(\frac{r}{r_1}\right) &+ \sum_{k=1}^{\infty}\left[\left(\frac{r}{r_1}\right)^k - \left(\frac{r}{r_1}\right)^{-k}\right](A_k\cos(k\theta) + B_k\sin(k\theta)) \\ +C_0 \ln\left(\frac{r}{r_2}\right) &+ \sum_{k=1}^{\infty}\left[\left(\frac{r}{r_2}\right)^k - \left(\frac{r}{r_2}\right)^{-k}\right](C_k\cos(k\theta) + D_k\sin(k\theta)), \end{aligned} \tag{5}$$

where the coefficients are determined as:

$$A_0 \ln\left(\frac{r_2}{r_1}\right) = \frac{1}{2\pi}\int_0^{2\pi} f_\alpha(t)\,dt, \qquad C_0 \ln\left(\frac{r_1}{r_2}\right) = \frac{1}{2\pi}\int_0^{2\pi} g(t)\,dt,$$

$$A_k\left[\left(\frac{r_2}{r_1}\right)^k - \left(\frac{r_2}{r_1}\right)^{-k}\right] = \frac{1}{\pi}\int_0^{2\pi} f_\alpha(t)\cos(kt)\,dt, \quad C_k\left[\left(\frac{r_1}{r_2}\right)^k - \left(\frac{r_1}{r_2}\right)^{-k}\right] = \frac{1}{\pi}\int_0^{2\pi} g(t)\cos(kt)\,dt,$$

$$B_k\left[\left(\frac{r_2}{r_1}\right)^k - \left(\frac{r_2}{r_1}\right)^{-k}\right] = \frac{1}{\pi}\int_0^{2\pi} f_\alpha(t)\sin(kt)\,dt, \quad D_k\left[\left(\frac{r_1}{r_2}\right)^k - \left(\frac{r_1}{r_2}\right)^{-k}\right] = \frac{1}{\pi}\int_0^{2\pi} g(t)\sin(kt)\,dt.$$

Differentiating Eq. (5) and applying the boundary condition in Eq. (3), we derive the Fredholm integral equation of the first kind:

$$\int_0^{2\pi} K(\theta, t) f(t)\,dt = H(\theta), \tag{6}$$

where $K(\theta, t)$ and $H(\theta)$ are defined as:

$$\begin{aligned} K(\theta,t) = \frac{1}{2\pi r_1}\left(\ln\left(\frac{r_2}{r_1}\right)\right)^{-1} &+ \sum_{k=1}^{\infty}\frac{2k}{\pi r_1}\left(\left(\frac{r_2}{r_1}\right)^k - \left(\frac{r_2}{r_1}\right)^{-k}\right)^{-1} \\ &\times (\cos(k\theta)\cos(kt) + \sin(k\theta)\sin(kt)), \end{aligned}$$

$$H(\theta) = h_1(\theta) - \int_0^{2\pi}\left[\sum_{k=1}^{\infty}\frac{k}{\pi}\left(\frac{r_1^{k-1}}{r_2^k} + \frac{r_1^{-k-1}}{r_2^{-k}}\right)\left(\left(\frac{r_1}{r_2}\right)^k - \left(\frac{r_1}{r_2}\right)^{-k}\right)^{-1}\right.$$

$$\times (\cos(k\theta)\cos(kt) + \sin(k\theta)\sin(kt)) + \frac{1}{2\pi r_1}\left(\ln\left(\frac{r_1}{r_2}\right)\right)^{-1}\Bigg] g(t)\,dt.$$

To address the ill-posed nature of Eq. (6), we introduce regularization, transforming it into a Fredholm equation of the second kind:

$$\alpha f_\alpha(\theta) + \int_0^{2\pi} K(\theta, t) f_\alpha(t)\,dt = H(\theta), \tag{7}$$

where α is a regularization parameter.

3 The Resolution Methods

The solution to the Fredholm equation is approximated using three key considerations. First, the series expansion of $K(\theta, t)$ and $H(\theta)$ is truncated to a finite order N. Second, the function f_α is projected onto a finite-dimensional subspace of $L^2([0, 2\pi])$ of dimension $2M$, represented by an orthogonal basis constructed from shifted normalized Legendre polynomials. Third, numerical integration, based on Haar wavelets, is applied to approximate the integrals involved. The influence of the parameters N and M will be analyzed later.

Numerical quadrature is employed to approximate the integrals in Eq. (7), defined as:

$$\int_0^{2\pi} g(t)\,dt \approx \frac{\pi}{M}\sum_{i=1}^{2M} g(x_i), \tag{8}$$

where the quadrature points x_i are given by:

$$x_i = \frac{\pi(i - 0.5)}{M}. \tag{9}$$

Substituting this approximation into Eq. (7) leads to:

$$\alpha f_\alpha(x) + \frac{\pi}{M}\sum_{i=1}^{2M} K(x, x_i) f_\alpha(x_i) \approx h(x). \tag{10}$$

The function $f_\alpha(x)$ is then approximated as:

$$f_\alpha(x) = \sum_{l=1}^{2M} a_l h_l(x), \tag{11}$$

where $h_l(x)$ are the shifted normalized Legendre polynomials defined on $[0, 2\pi]$. These are constructed from classical Legendre polynomials $L_j(x)$, which satisfy the recurrence relation:

$$L_0(x) = 1, \quad L_1(x) = x,$$

$$L_j(x) = \frac{(2j-1)xL_{j-1}(x) - (j-1)L_{j-2}(x)}{j}, \quad j \geq 2.$$

To shift the interval $[a, b]$ to $[-1, 1]$, the transformation:

$$x' = \frac{2x - (a+b)}{b-a},$$

is applied, where $x' \in [-1, 1]$ and $x \in [a, b]$. The normalized shifted Legendre polynomials $h_j(x)$ on $[a, b]$ are then expressed as:

$$h_j(x) = \sqrt{\frac{2j+1}{b-a}} L_j(x').$$

Using this representation, Eq. (10) becomes:

$$\alpha \sum_{l=1}^{2M} a_l h_l(x) + \frac{\pi}{M} \sum_{l=1}^{2M} \sum_{i=1}^{2M} K(x, x_i) a_l h_l(x_i) = h(x). \tag{12}$$

At specific points x_p in $[0, 2\pi]$, we derive the following linear system:

$$A\mathbf{a} = h, \tag{13}$$

where the matrix A, the coefficient vector $\mathbf{a}$, and the right-hand side h are defined as:

$$A_{pl} = \alpha h_l(x_p) + \frac{\pi}{M} \sum_{i=1}^{2M} K(x_p, x_i) h_l(x_i), \quad p, l = 1, \ldots, 2M, \tag{14}$$

$$\mathbf{a} = \left(a_1, \cdots, a_{2M}\right)^T, \tag{15}$$

$$h = \left(h(x_1), \cdots, h(x_{2M})\right)^T. \tag{16}$$

Letting $H_{i,j} = h_i(x_j)$, the matrix A can be expressed as:

$$A = \alpha I_{2M} + \frac{\pi}{M} K H^T,$$

where I_{2M} is the identity matrix, and H^T denotes the transpose of H and K is the matrix defined by $K_{p,i} = K(x_p, x_i)$. Multiplying both sides of Eq. (13) by A^T, we obtain the system:

$$A^T A \, \mathbf{a} = A^T h,$$

which is solved using the conjugate gradient method. Finally, the solution f_α at points x_p is computed as:

$$F = H^T \mathbf{a},$$

where $F = (f_\alpha(x_1), \cdots, f_\alpha(x_{2M}))^T$.

4 Numerical Results

In this section, we calculate all coefficients of the regularized solution, setting $r_1 = 2$ and $r_2 = 1$.

4.1 *Example 1*

We consider the following data:

$$\begin{cases} g(\theta) = r_1^2 \cos(2\theta), \\ f(\theta) = r_2^2 \cos(2\theta), \\ h_1(\theta) = 2r_1 \cos(2\theta). \end{cases}$$

The exact solution is $u(r, \theta) = r^2 \cos(2\theta)$.

For this example, we derive an analytical expression for $H(\theta)$ since the integrals can be computed as:

$$\int_0^{2\pi} g(t)\, dt = r_1^2 \int_0^{2\pi} \cos(2\theta)\, d\theta = 0,$$

and by the orthogonality of $\cos(k\theta)$, we have:

$$\int_0^{2\pi} \cos(k\theta) \cos(2\theta)\, d\theta = \pi \delta_{k,2},$$

where $\delta_{k,2}$ is the Kronecker delta. Additionally:

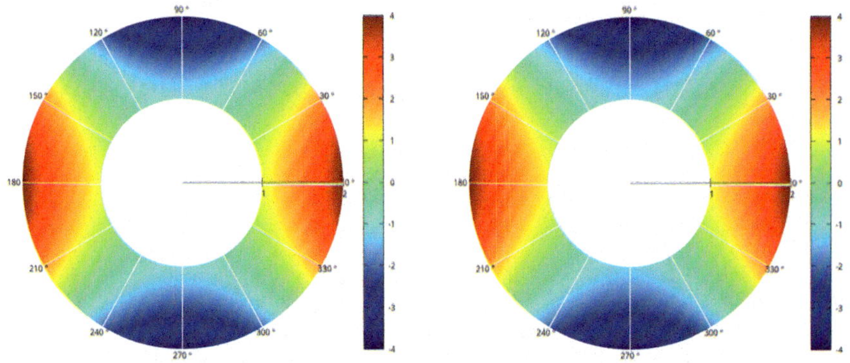

(a) Exact solution $u(r, \theta) = r^2 \cos(2\theta)$ (b) Regularized solution u_α

Fig. 1 Exact and Regularized solutions

Table 1 Error evolution with respect to N

Value of N	5	10	15	20
Error	0.19004	0.017047	0.0160407	0.014020

$$\int_0^{2\pi} \sin(k\theta)\cos(2\theta)\, d\theta = 0 \quad \forall k.$$

Hence, $H(\theta)$ is given by:

$$H(\theta) = 2r_1 \cos(2\theta) - r_1^2 \cos(2\theta) \left[\frac{2r_1}{r_2^2} + \frac{2r_1^{-3}}{r_2^{-2}} \right] \left[\left(\frac{r_1}{r_2}\right)^2 - \left(\frac{r_1}{r_2}\right)^{-2} \right]^{-1}.$$

The system is solved numerically as described in Sect. 3, yielding F. Using these coefficients, we compute the regularized solution u_α. The solutions are visualized in Fig. 1, comparing the regularized and exact solutions.

The regularized solution closely matches the exact solution, as shown in Fig. 1.

Influence of Parameters M, N, and α We analyze the influence of the parameter N by setting $M = 5$ and $\alpha = 0.01$. The numerical integration method employed is used is that defined in (8)–(9). Table 1 shows the error evolution as N increases. The error decreases significantly until $N = 7$, beyond which further increases in N have minimal impact.

Regarding M, we observe that a small M achieves good precision for this example, as we are approximating a polynomial function in a polynomial basis.

To study the effect of the regularization parameter α, we fix $N = 5$ and $M = 5$, varying α from 0 to 10^{-3}. Table 2 summarizes the error for different α values. The error is large for $\alpha = 0$, confirming the problem's ill-posed nature and the necessity of regularization.

Table 2 Error evolution with respect to α

Value of α	0	0.1	0.05	0.01	0.005	0.001	0.0001
Error	0.646371	0.229505	0.102882	0.019004	0.009664	0.002118	0.000424

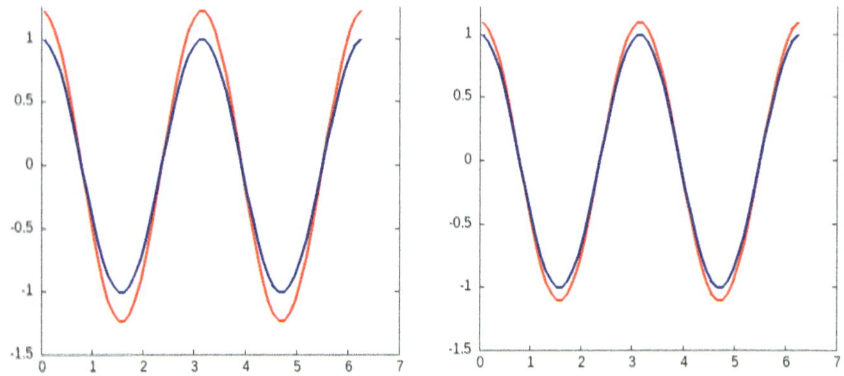

(a) Solution for $\alpha = 0.1$ (b) Solution for $\alpha = 0.05$

Fig. 2 Regularized solutions for $\alpha = 0.1$ and $\alpha = 0.05$

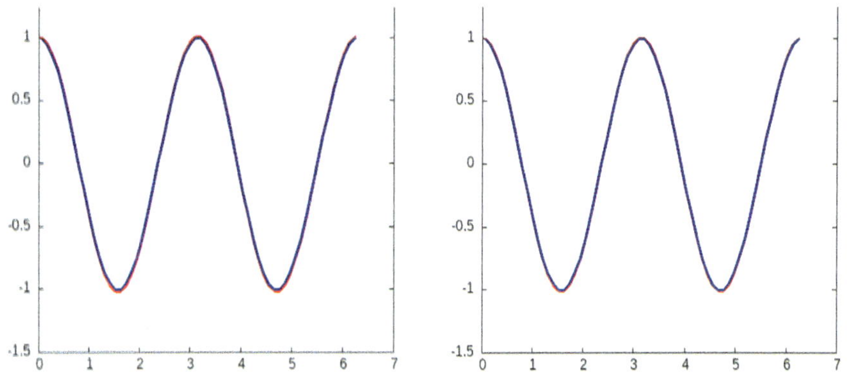

(a) Solution for $\alpha = 0.01$ (b) Solution for $\alpha = 0.005$

Fig. 3 Regularized solutions for $\alpha = 0.01$ and $\alpha = 0.005$

Figures 2 and 3 illustrate the solutions for different α values. As α decreases, the regularized solution approaches the exact solution.

4.2 Second Example

For the second example, we take the following data:

$$\begin{cases} g(\theta) = \cos(r_1 \sin\theta) e^{r_1 \cos\theta}, \\ f(\theta) = \cos(r_2 \sin\theta) e^{r_2 \cos\theta}, \\ h_1(\theta) = e^{r_1 \cos\theta} (\cos\theta \cos(r_1 \sin\theta) - \sin\theta \sin(r_1 \sin\theta)) . \end{cases}$$

The exact solution is given by:

$$u(r,\theta) = \cos(r \sin\theta) e^{r \cos\theta}.$$

This example presents additional computational challenges. Unlike the previous case, $H(\theta)$ cannot be determined analytically and requires numerical quadrature methods for approximation. The computation of $H(\theta)$ consists of two parts:

The first term is approximated as:

$$A = \int_0^{2\pi} \frac{1}{2\pi r_1} \left(\ln\left(\frac{r_1}{r_2}\right)\right)^{-1} g(t)\, dt \approx \frac{1}{2M r_1} \left(\ln\left(\frac{r_1}{r_2}\right)\right)^{-1} \sum_{i=1}^{2M} g(x_i).$$

The second term involves an infinite series, truncated after N terms:

$$B \approx \sum_{k=1}^{N} \frac{1}{M} \left(\frac{k r_1^{k-1}}{r_2^k} + \frac{k r_1^{-k-1}}{r_2^{-k}} \right) \left(\left(\frac{r_1}{r_2}\right)^k - \left(\frac{r_1}{r_2}\right)^{-k} \right)^{-1}$$
$$\times \sum_{i=1}^{2M} (\cos(k\theta)\cos(kx_i) + \sin(k\theta)\sin(kx_i)) g(x_i).$$

The value of $H(\theta)$ is obtained as:

$$H(\theta) = h_1(\theta) - A - B.$$

The regularized solution u^α is computed by solving the linear system described in Sect. 3.

Influence of Parameters M, N, and α As in Example 1, we examine the influence of the parameters M, N, and α on the accuracy of the regularized solution. For fixed M and α, the error decreases with N until stagnation, as the series terms become negligible.

For M, the observations are consistent with those from Example 1. We use $M = 10$ and study the effect of varying α from 0.1 to 10^{-3}. Table 3 shows that smaller values of α lead to more accurate results, demonstrating the importance of the regularization parameter in controlling solution precision.

Table 3 Error evolution with respect to α

Value of α	0.1	0.05	0.01	0.005	0.001	0.0001
Error	0.525342	0.245316	0.044726	0.022437	0.004500	0.000475

5 Conclusion

In this study, we investigated the approximation the solution of the inverse Cauchy problem for Laplace's equation, which arises in nondestructive testing of conductive materials and emphasize the challenges posed by the problem's ill-posed nature. We developed a semi-analytical approach based on transforming the inverse Cauchy problem into a Fredholm integral equation and using Legendre polynomial expansion.

We examined the impact of key parameters, including the number of Fourier terms N, quadrature points M, and the regularization parameter α. Through numerical experiments, we analyzed the influence of these parameters on approximation accuracy and error behavior, providing insights into how parameter selection can achieve a balance between computational efficiency and solution precision.

Future work could extend this analysis to more complex problem or consider adaptive methods for selecting N and α based on the specific characteristics of the input data which further enhance the applicability of this approach in real-world scenarios. Overall, the method and insights provided here contribute meaningfully to the field of numerical analysis and can serve as a reference for solving similar approximation problems.

Acknowledgements This work was developed during the visit of Sudad Musa Rashid and Gulnar Wasim Sadiq, at the Laboratoire de Mathématiques Jean Leray, Nantes Université (France), CNRS UMR 6629. This scientific stay was partially funded by Le Service de Coopérationd'Action Culturelle (SCAC), Ambassade de France en Irak.

References

1. Jourhmane, M., Nachaoui, A.: An alternating method for an inverse cauchy problem. Numer. Algorithms **21**(1), 247–260 (1999)
2. Nachaoui, A.: Numerical linear algebra for reconstruction inverse problems. J. Comput. Appl. Math. **162**, 147–164 (2004)
3. Regińska, T., Regiński, K.: Approximate solution of a cauchy problem for the helmholtz equation. Inverse Prob. **22**(3), 975–989 (2006)
4. Arsenashvili, A., Nachaoui, A., Tadumadze, T.: On approximate solution of an inverse problem for linear delay differential equations. Bull. Georgian Natl. Acad. Sci. (N.S.) **2**(2), 24–28 (2008)
5. Ellabib, A., Nachaoui, A.: An iterative approach to the solution of an inverse problem in linear elasticity. Math. Comput. Simul. **77**(2–3), 189–201 (2008)
6. Fu, C.L., Feng, X.L., Qian, Z.: The fourier regularization for solving the cauchy problem for the helmholtz equation. Appl. Numer. Math. **59**(10), 2625–2640 (2009)

7. Qian, A.L., Xiong, X.T., Wu, Y.-J.: On a quasi-reversibility regularization method for a cauchy problem of the helmholtz equation. J. Comput. Appl. Math. **233**(8), 1969–1979 (2010)
8. Aboud, F., Nachaoui, A., Nachaoui, M.: On the approximation of a cauchy problem in a non-homogeneous medium. J. Phys.: Conf. Ser. **1743**(1), 012003 (2021)
9. Berdawood, K.A., Nachaoui, A., Nachaoui, M., Aboud, F.: An effective relaxed alternating procedure for cauchy problem connected with helmholtz equation. Numer. Methods Part. Diff. Equ. **39**(3), 1888–1914 (2022)
10. Grunau, H.-C., Miyake, N., Okabe, S.: Positivity of solutions to the cauchy problem for linear and semilinear biharmonic heat equations. Adv. Nonlinear Anal. **10**(1), 353–370 (2021)
11. Nachaoui, A., Nachaoui, M., Chakib, A., Hilal, M.: Some novel numerical techniques for an inverse cauchy problem. J. Comput. Appl. Math. **381** (113030) (2021)
12. Ellabib, A., Nachaoui, A., Ousaadane, A.: Convergence study and regularizing property of a modified robin-robin method for the cauchy problem in linear elasticity. Inverse Prob. **38**(7), 075007 (2022)
13. Nachaoui, A., Gontier, L.: Analytical Solutions for Inverse Cauchy Problems: A Comparative Study. Springer Proceedings in Mathematics & Statistics, vol. 498. Springer, Cham (2025)
14. Hadri, A., Laghrib, A., Nachaoui, A., Nachaoui, M.: On the resolution of the non-smooth inverse cauchy problem by the primal-dual method, Numerical Algorithms (2024). https://doi.org/10.1007/s11075-024-01950-5
15. Nachaoui, A., Nachaoui, M., Tadumadze, T.: Meshless Methods to Noninvasively Calculate Neurocortical Potentials from Potentials Measured at the Scalp Surface. Springer Proceedings in Mathematics & Statistics, vol. 428, pp. 99–117. Springer, Cham (2023)
16. Ellabib, A., Nachaoui, A.: On the numerical solution of a free boundary identification problem. Inverse Probl. Eng. **9**(3), 235–260 (2001)
17. Nachaoui, A.: An improved implementation of an iterative method in boundary identification problems. Numer. Algorithms **33**(1–4), 381–398 (2003)
18. Chakib, A., Nachaoui, A., Zeghal, A.: A shape optimization approach for an inverse heat source problem. Int. J. Nonlinear Sci. **12**(1), 78–84 (2012)
19. Isakov, V.: Inverse problems for partial differential equations. Applied Mathematical Sciences, vol. 127. Springer, Cham (2017)
20. Nachaoui, M., Nachaoui, A., Tadumadze, T.: On the numerical approximation of some inverse problems governed by nonlinear delay differential equation. RAIRO Oper. Res. **56**(3), 1553–1569 (2022)
21. Lavrentiev, M.M.: Some Improperly Posed Problems of Mathematical Physics. Springer Science & Business Media (2013)
22. Huang, C.H., Chen, W.C.: A three-dimensional inverse forced convection problem in estimating surface heat flux by conjugate gradient method. Int. J. Heat Mass Transf. **43**(17), 3171–3181 (2000)
23. Wang, F., Gu, Y., Qu, W., Zhang, C.: Localized boundary knot method and its application to large-scale acoustic problems. Comput. Methods Appl. Mech. Eng. **361**, 112729 (2020)
24. Ellabib, A., Nachaoui, A., Ousaadane, A.: Mathematical analysis and simulation of fixed point formulation of cauchy problem in linear elasticity. Math. Comput. Simul. **187**, 231–247 (2021)
25. Nachaoui, A., Laghrib, A., Hakim, M.: Mathematical control and numerical applications. Springer Proceedings in Mathematics & Statistics, vol. 372. Springer, Cham (2021)
26. Nachaoui, A., Salih, H.W.: An analytical solution for the nonlinear inverse cauchy problem. Adv. Math. Models Appl. **6**(3), 191–206 (2021)
27. Nachaoui, A., Rashid, S.M.: A mesh free wavelet method to solve the cauchy problem for the helmholtz equation. Springer Proceedings in Mathematics & Statistics, vol. 428, pp. 77–98. Springer Cham (2023)
28. Ellabib, A., Nachaoui, A., Ousaadane, A.: A relaxed robin-robin algorithm for solving a cauchy problem in linear elasticity. J. Appl. Math. Comput. **71**, 1721–1741 (2025). https://doi.org/10.1007/s12190-024-02295-x
29. Nachaoui, A.: Cauchy's problem for the modified biharmonic equation : Derivation and decoupled iterative methods. Springer Proceedings in Mathematics & Statistics, vol. 498. Springer, Cham (2025)

30. Vasil'ev, V.V.P.V.I., Kardashevsky, A.M., Prokopev, G.A.: Iterative solution of the inverse cauchy problem for an elliptic equation by the conjugate gradient method. AIP Conf. Proc. **1895**(1), 110009 (2017)
31. Berdawood, K., Nachaoui, A., Nachaoui, M.: An accelerated alternating iterative algorithm for data completion problems connected with helmholtz equationn. Stat. Opt. & Inf. Comput. **11**(1), 2–21 (2023). https://doi.org/10.1002/num.22793
32. Nachaoui, A., Aboud, F., Nachaoui, M.: Acceleration of the kmf algorithm convergence to solve the cauchy problem for poisson's equation. Springer Proceedings in Mathematics & Statistics, vol. 372, pp. 43–57. Springer, Cham (2021)
33. Liu, J.C., Wei, T.: A quasi-reversibility regularization method for an inverse heat conduction problem without initial data. Appl. Math. Comput. **219**(23), 10866–10881 (2013)
34. Lattès, R., Lions, J.: The method of quasi-reversibility: applications to partial differential equations, Modern Analytic and Computational Methods in Science and Mathematics. American Elsevier Publishing Co., Inc, New York (1969)
35. Cheng, X., Gong, R., Han, W.: A coupled complex boundary method for the cauchy problem. Inverse Probl. Sci. Eng. **24**(9), 1510–1527 (2016)
36. Belytschko, T., Krongauz, Y., Organ, D., Fleming, M., Krysl, P.: Meshless methods: an overview and recent developments. Comput. Methods Appl. Mech. Engrg. **139**, 3–47 (1996)
37. Fairweather, G., Karageorghis, A.: The method of fundamental solutions for elliptic boundary value problems. Adv. Comput. Math. **9**, 69–95 (1998)
38. Shen, S., Atluri, S.N.: The Meshless Local Petrov-Galerkin (MLPG) Method (2002)
39. Jin, B., Zheng, Y.: A meshless method for some inverse problems associated with the helmholtz equation. Comput. Methods Appl. Mech. Eng. **195**(19–22), 2270–2288 (2006)
40. Wei, T., Hon, Y.C., Ling, L.: Method of fundamental solutions with regularization techniques for cauchy problems of elliptic operators. Eng. Anal. Boundary Elem. **31**(4), 373–385 (2007)
41. Hua, Q., Gu, Y., Qu, W., Chen, W., Zhang, C.: A meshless generalized finite difference method for inverse cauchy problems associated with three-dimensional inhomogeneous helmholtz-type equations. Eng. Anal. Bound. Elem. **82**, 162–171 (2017)
42. Yue, X., Wang, F., Hua, Q., Qiu, X.: A novel space-time meshless method for nonhomogeneous convection-diffusion equations with variable coefficients. Appl. Math. Lett. **92**, 144–150 (2019)
43. Rasheed, S.M., Nachaoui, A., Hama, M.F., Jabbar, A.K.: Regularized and preconditioned conjugate gradient like-methods methods for polynomial approximation of an inverse cauchy problem. Adv. Math. Models & Appl. **6**(2), 89–105 (2021)
44. Rashid, S.M., Nachaoui, A.: A haar wavelets-based direct reconstruction method for the cauchy problem of the poisson equation. Discrete Contin. Dyn. Syst. Ser. S **18**(1), 148–168 (2025)
45. Fang, W., Lu, M.: A fast collocation method for an inverse boundary value problem. Int. J. Numer. Meth. Eng. **21**, 1563–1585 (2004)
46. Liu, C.S., Atluri, S.N.: A highly accurate technique for interpolations using very high-order polynomials, and its applications to some ill-posed linear problems. Comput. Model. Eng. & Sci. **43**(1), 253–276 (2009)
47. Reddy, G.M.M., Nanda, P., Vynnycky, M., Cuminato, J.A.: An adaptive boundary algorithm for the reconstruction of boundary and initial data using the method of fundamental solutions for the inverse cauchy-stefan problem. Comput. Appl. Math. **40**(3) (2021)
48. Aboud, F., Jameel, I.T., Hasan, A.F., Mostafa, B.K., Nachaoui, A.: Polynomial approximation of an inverse cauchy problem for helmholtz type equations. Adv. Math. Models & Appl. **7**(3), 306–322 (2022)
49. Nachaoui, A., Sadiq, G.W.: Identification of a space dependent force in a hyperbolic equation using a polynomial expansion. Springer Proceedings in Mathematics & Statistics, vol. 498. Springer, Cham (2025)
50. Nachaoui, A., Sadiq, G.W.: Investigation of an inverse problem in heat propagation, To appear (2025)

Analytical Solutions for Inverse Cauchy Problems: A Comparative Study

Abdeljalil Nachaoui and Lucien Gontier

Abstract This paper investigates two analytical methods for solving the inverse Cauchy problem for Laplace's equation, focusing on the reconstruction of the solution $u(r, \theta)$ from boundary data. The first method employs a regularized solution approach based on solving a Fredholm integral equation of the first kind, while the second utilizes the homotopy perturbation method (HPM), a series-based analytical technique. Both methods are tested on a benchmark problem with a known exact solution, allowing a comprehensive comparison of their accuracy, stability, and computational efficiency.

Keywords Inverse Cauchy problem · Analytical approach · Fredholm integral equation · Homotopy perturbation method

1 Introduction

Electrostatic imaging in nondestructive testing of metallic disks leads to an inverse boundary value problem involving the Laplace equation in two dimensions. To detect the unknown shape of an inclusion within a conductive material, one must impose additional Cauchy data-such as voltage and current-on the accessible external boundary, creating an inverse problem based on boundary data [1–21]. This problem is well-known for being highly ill-posed [22–24].

To model this mathematically, consider Cauchy data $u(x, y)$ and $u_n(x, y)$ at a point $(x, y) \in \mathbb{R}^2$, where $u_n(x, y)$ denotes the normal derivative of $u(x, y)$ on the outer circle Γ_1 of radius r_1 in an annular domain Ω. We seek to solve the Laplace equation $\Delta u(x, y) = 0$ in two dimensions to determine the unknown function $u(x, y)$ on an internal circle Γ_2 of radius $r_2 < r_1$. This model is applicable to inverse electro-

A. Nachaoui (✉) · L. Gontier
Laboratoire de Mathématiques Jean Leray, Nantes Université, Nantes, France
e-mail: Abdeljalil.Nachaoui@univ-nantes.fr

L. Gontier
e-mail: lucien.gontier@etu.univ-nantes.fr

A. Laghrib et al. (eds.), *Computational Methods for Inverse Problems and Applications*, Springer Proceedings in Mathematics & Statistics 498,
https://doi.org/10.1007/978-3-031-89498-5_5

static imaging problems in human electroencephalography [2, 25–27]. Specifically, the problem consists of solving a disk-based equation using overspecified Cauchy boundary data. Several works have been published in recent decades concerning the resolution of this problem [13, 17, 19, 20, 28–31]. Among these methods, the so-called meshless methods, which replace meshing with a set of arbitrarily distributed nodes, have been proposed in the following [5, 32–45].

1.1 Problem Studied

Consider an annular region with radii r_2 and r_1, where $r_2 < r_1$. The Cauchy problem is as follows:

$$\begin{cases} \Delta u(r,\theta) = u_{rr}(r,\theta) + \frac{1}{r}u_r(r,\theta) + \frac{1}{r^2}u_{\theta\theta}(r,\theta) = 0, & r_2 < r < r_1,\ 0 \le \theta \le 2\pi \\ u(r_1,\theta) = g(\theta), & 0 \le \theta \le 2\pi \\ u_r(r_1,\theta) = h_1(\theta), & 0 \le \theta \le 2\pi, \end{cases} \tag{1}$$

where $u_x = \frac{\partial u}{\partial x}$ for $x = r$ or $x = \theta$.

1.2 The Analytical Regularized Solution Approach

The above Cauchy problem can be regularized by converting it to a Fredholm integral equation of the second kind [46–48]. Using kernel function separation and eigenfunction expansion techniques, the regularized solution $u^\alpha(r,\theta)$ of the integral equation takes the form:

$$u^\alpha(r,\theta) = \frac{1}{2}(a_0^\alpha + b_0^\alpha \ln(r)) + \sum_{k=1}^{\infty}\left((a_k^\alpha r^k + b_k^\alpha r^{-k})\cos(k\theta) + (c_k^\alpha r^k + d_k^\alpha r^{-k})\sin(k\theta)\right) \tag{2}$$

where the coefficients are given by:

$$\begin{aligned}
a_0^\alpha &= \frac{r_1\ln(r_1)}{\pi[1+\alpha r_1(\ln(r_1)-\ln(r_2))]}\int_0^{2\pi} h(\xi)\,d\xi - \frac{\ln(r_2)}{\pi(\ln(r_1)-\ln(r_2))}\int_0^{2\pi} g(\xi)\,d\xi,\\
b_0^\alpha &= \frac{1}{\pi(\ln(r_1)-\ln(r_2))}\int_0^{2\pi} g(\xi)\,d\xi - \frac{r_1}{\pi[1+\alpha r_1(\ln(r_1)-\ln(r_2))]}\int_0^{2\pi} h(\xi)\,d\xi,\\
a_k^\alpha &= \frac{e_k}{r_2^k}\int_0^{2\pi} g(\xi)\cos(k\xi)\,d\xi - \frac{e_k}{r_1^k(\alpha+\pi B_k)}\int_0^{2\pi} h(\xi)\cos(k\xi)\,d\xi,\\
b_k^\alpha &= \frac{e_k r_1^k}{\alpha+\pi B_k}\int_0^{2\pi} h(\xi)\cos(k\xi)\,d\xi - e_k r_2^k\int_0^{2\pi} g(\xi)\cos(k\xi)\,d\xi,
\end{aligned}$$

$$c_k^\alpha = \frac{e_k}{r_2^k}\int_0^{2\pi} g(\xi)\sin(k\xi)\,d\xi - \frac{e_k}{r_1^k(\alpha+\pi B_k)}\int_0^{2\pi} h(\xi)\sin(k\xi)\,d\xi,$$
$$d_k^\alpha = \frac{e_k r_1^k}{\alpha+\pi B_k}\int_0^{2\pi} h(\xi)\sin(k\xi)\,d\xi - e_k r_2^k\int_0^{2\pi} g(\xi)\sin(k\xi)\,d\xi,$$

with

$$B_k := 2r_1^{-1}ke_k, \quad e_k = \frac{1}{\pi\left(\left(\frac{r_1}{r_2}\right)^k - \left(\frac{r_2}{r_1}\right)^k\right)}.$$

Define

$$h(\theta) := -h_1(\theta) + \frac{1}{2r_1\pi(\ln(r_1)-\ln(r_2))}\int_0^{2\pi} g(\xi)\,d\xi$$
$$+\sum_{k=1}^{\infty}\left(A_k\int_0^{2\pi}(g(\xi)\cos(k\xi)\cos(k\theta)+g(\xi)\sin(k\xi)\sin(k\theta))\,d\xi\right),$$
$$A_k := kr_1^{k-1}e_k r_2^{-k} + kr_1^{-k-1}e_k r_2^k.$$

We assume $0 < r_2 < r_1$.

In this work we will follow the idea of [49] by applying the homotopy perturbation method. The proposed method, which does not require any small parameters in the equations, easily eliminates the drawback of the Thikhonov regularization approximation used in the analytical method developed by [46], which requires an appropriate choice of the regularization parameter in order to lead to ideal results. However, an inappropriate choice of this parameter results in bad effects, sometimes seriously.

In the next section, we outline the steps required to derive an analytical solution using the homotopy perturbation method. Following that, in the numerical results section, we compare the outcomes of both methods for determining an internal boundary and the Robin coefficient.

2 Deriving a Homotopy Perturbation Based Solution

To derive an analytical solution for the inverse Cauchy problem using the homotopy perturbation method (HPM), we can proceed as follows:

Step 1: Reformulate the problem

We decompose our differential operator Δ according to the following component.

$$\Delta u = \underbrace{r^2u_{rr} + ru_r}_{R(u)} + \underbrace{u_{\theta\theta}}_{T(u)}.$$

Step 2: Construct the Homotopy Equation
We define an homotopy $v((r, \theta), p) : \Omega \times [0, 1] \to \mathbb{R}$, associated with the homotopy parameter p such that :

$$\mathcal{H}(v, p) = (1 - p)[R(v) - R(u_0)] + p[R(v) + T(v)] = 0, \tag{3}$$

with u_0 an initial approximation of the problem satisfying the Cauchy boundary conditions.
From the previous equation, we have :

$$\mathcal{H}(v, 0) = R(v) - R(u_0) = 0,$$
$$\mathcal{H}(v, 1) = R(v) + T(v) = \Delta v = 0,$$

the changing process of p from zero to unity is just that of $v(r, p)$ from $u_0(r)$ to $u(r)$.
Step 3: Expending the solution in term of p
We assume that for a small parameter p, we can expand the solution v of Eq. (3) as a power series in p in the following form:

$$v = v^0 + pv^1 + p^2 v^2 + \cdots \tag{4}$$

Then, for each order we can get an equation, solve it and go to the next order.
For the order p^0, we get :

$$v = v^0 + O(p^1),$$

inserting it in (3), we get :

$$R(v^0) - R(u_0) + O(p^1) = 0.$$

Since we took u_0 an initial approximation of the problem, we get v^0 solution of

$$r^2 u^0_{rr} + r u^0_r = 0. \tag{5}$$

For the order p^1, we get :

$$v = v^0 + pv^1 + O(p^2).$$

Again, inserting this equation in (3) and using the previous order, we get :

$$pR(v^1) + p[R(v^0) + T(v^0)] + O(p^2) = 0.$$

This lead us to the following equation :

$$r^2 v^1_{rr} + r v^1_r = -v^0_{\theta\theta},$$

By the same manipulation, we get for the order p^k, $k \in \mathbb{N}$:

$$r^2 v^k_{rr} + r v^k_r = -v^{k-1}_{\theta\theta}.$$

Step 4: Solve equation Sequentially
To solve the above equation, we impose the boundaries conditions as follow:

$$p^0 : \begin{cases} v^0(r_1, \theta) = g(\theta), \\ v^0_r(r_1, \theta) = h_1(\theta), \end{cases}$$

$$k \geq 1,\ p^k : \begin{cases} v^k(r_1, \theta) = 0, \\ v^k_r(r_1, \theta) = 0 \end{cases}$$

The solutions of the above equations can be obtained analytically. We find solutions of the two first equations, the others can be obtained by induction. For the sake of simplification and without any restriction, we assume in the following that $r_1 = 1$
• For v^0, we begin by solving the first equation in terms of v^0_r, we use the boundary condition $v^0_r(r_1, \theta) = h_1(\theta)$ and obtain:

$$v^0_r(r, \theta) = \frac{h_1(\theta)}{r}, \qquad \forall (r, \theta) \in \Omega$$

Then integrating on $[r, r_1]$ and using boundary condition $v^0(r_1, \theta) = g(\theta)$ we get :

$$v^0(r, \theta) = g(\theta) + h_1(\theta)\ln(r), \qquad \forall (r, \theta) \in \Omega$$

• The first-order approximation of Eq. (1), can be obtained as follows: note that the equation verified by v^1_r is a first order differential equation with constant second member $v^0_{\theta\theta}$, we therefore find the solution of the homogeneous equation as for order 0, then use the technique of variation of constants and the boundary conditions to obtain v^1_r, the solution of the non non-homogeneous equation:

$$v^1_r(r, \theta) = -g''(\theta)\frac{\ln(r)}{r} - h''_1(\theta)\frac{(\ln(r))^2}{2r}, \qquad \forall (r, \theta) \in \Omega$$

Then integrating on $[r, r1]$, and using the boundary condition $v^1(r_1, \theta) = 0$:

$$v^1(r, \theta) = -\frac{g^{(2)}(\theta)}{2}(\ln r)^2 - \frac{h^{(2)}_1(\theta)}{3!}(\ln r)^3, \qquad \forall (r, \theta) \in \Omega,$$

where $g^{(2)}$ and $h^{(2)}_1$ dénote the second derivative of g and h_1 respectively.
Solutions for high orders are done iteratively in the same way as the previous solution, we then obtain:

$$
\begin{aligned}
v^0(r,\theta) &= g(\theta) + h_1(\theta)\ln r,\\
v^1(r,\theta) &= -\frac{g^{(2)}(\theta)}{2}(\ln r)^2 - \frac{h_1^{(2)}(\theta)}{3!}(\ln r)^3,\\
v^2(r,\theta) &= \frac{g^{(4)}(\theta)}{4!}(\ln r)^4 + \frac{h_1^{(4)}(\theta)}{5!}(\ln r)^5,\\
&\vdots\\
v^k(r,\theta) &= (-1)^k\Big[\frac{g^{(2k)}(\theta)}{2k!}(\ln r)^{2k} + \frac{h_1^{(2k)}(\theta)}{(2k+1)!}(\ln r)^{2k+1}\Big],\\
&\vdots
\end{aligned}
$$

where $g^{(k)}$ and $h_1^{(k)}$ are the k-th derivative with respect of θ of g and h_1 respectively.

Step 5: Sum the Series Solution

Setting $p = 1$ results in the approximate solution of the Cauchy problem (1)

$$
u = \lim_{p\to 1} v = v^0 + v^1 + v^2 + \cdots,
$$

From the previous computation, we get :

$$
u(r,\theta) = \sum_{n=0}^{\infty} \frac{(-1)^n g^{(2n)}(\ln r)^{2n}}{(2n)!} + \sum_{n=0}^{\infty} \frac{(-1)^n h_1^{(2n)}(\ln r)^{2n+1}}{(2n+1)!} \tag{6}
$$

Step 6: Verify the Solution and Adjust if Necessary

Ensure it satisfies the boundary conditions. Set a parameter M for the truncations of the series. Test the approximated solution. Then, if necessary, add higher-order terms for improved accuracy.

Using this method, we obtain an approximate solution for $u(r,\theta)$ that meets the boundary conditions, providing an analytical form for the inverse Cauchy problem. This is beneficial, as it allows us to simply truncate the series to obtain the solution, without needing to compute integrals involving the boundary conditions, as is the case in the Regularized Solution Approach.

3 Numerical Results and Discussion

To assess the effectiveness of the two analytical methods, we consider a case where the exact solution of the Cauchy problem (1) is known:

$$g(\theta) = r_1^2 \cos(2\theta), \quad h_1(\theta) = 2r_1 \cos(2\theta),$$

corresponding to the exact solution $u(r, \theta) = r^2 \cos(2\theta) = x^2 - y^2$.

Substituting these values into the expression for $h(\theta)$, and using the orthogonality of the functions $\cos(k\xi)$ and $\sin(k\theta)$ along with their integration over $(0, 2\pi)$, we find that $h(\theta)$ simplifies to:

$$h(\theta) = (\pi A_2 r_1^2 - 2r_1) \cos(2\theta) \tag{7}$$

By substituting $g(\theta)$ and $h(\theta)$ into the definitions of a_k^α, b_k^α, c_k^α, and d_k^α, and applying the aforementioned properties, it follows that only the coefficients a_2^α and b_2^α are non-zero. These coefficients are given by:

$$a_2^\alpha = \frac{\pi r_1^2 e_2}{r_2^2} - \frac{\pi e_2(A_2 \pi r_1^2 - 2r_1)}{r_1^2(\alpha + \pi B_2)}, \quad \text{and} \quad b_2^\alpha = \frac{e_2 r_1^2 \pi (A_2 \pi r_1^2 - 2r_1)}{\alpha + \pi B_2} - e_2 r_2^2 \pi r_1^2.$$

The general analytical solution (2) simplifies to:

$$u^\alpha(r, \theta) = (a_2^\alpha r^2 + b_2^\alpha r^{-2}) \cos(2\theta)$$

In our numerical approach, we approximate integrals using the midpoint method.

We define: N_r: number of points in the $r \in [r_2, r_1]$ mesh, N_θ: number of points in the $\theta \in [0, 2\pi]$ mesh, m: upper limit of the sum in the analytical solution, and n: number of subdivisions in the numerical integration.

The objective is to determine how accurately each method approximates the exact solution and to compare the computational efficiency and stability of each approach, especially when handling Cauchy problems that tend to be ill-posed.

For the remainder of this discussion, we set $N_\theta = 1000$, $N_r = 1000$, with $r_1 = 1$ and $r_2 = 0.5$.

3.1 *Regularized Solution Approach*

The regularized analytical method was tested for varying regularization parameters α. Without noise in the boundary condition, the solution converges to the exact solution as $\alpha \to 0$, as shown in Fig. 1.

However, the method demonstrates a significant sensitivity to noise. When a noise level of 10^{-2} was added to g, the error increased drastically for small values of α (For example, the error corresponding to $\alpha = 10^{-8}$ is multiplied by 10^6), as shown in Fig. 2. This behavior illustrates that while the method can produce accurate results in ideal conditions, the choice of α is critical for stability, especially in noisy scenarios.

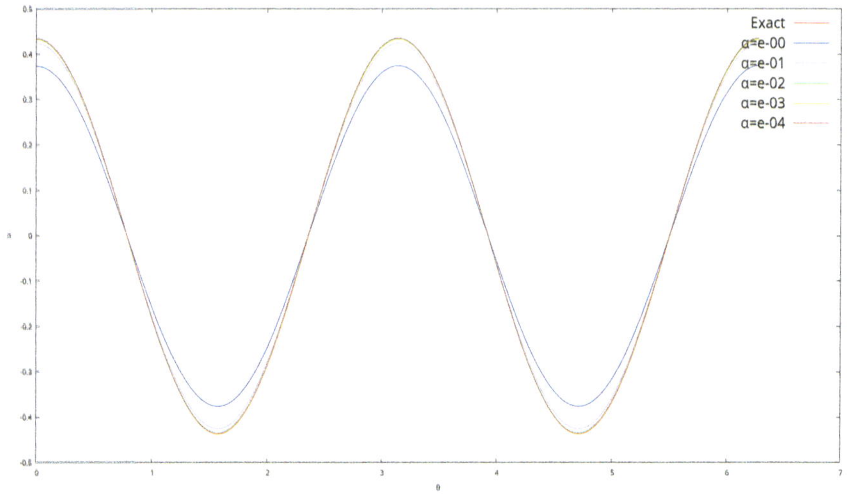

Fig. 1 Solutions for different regularization values

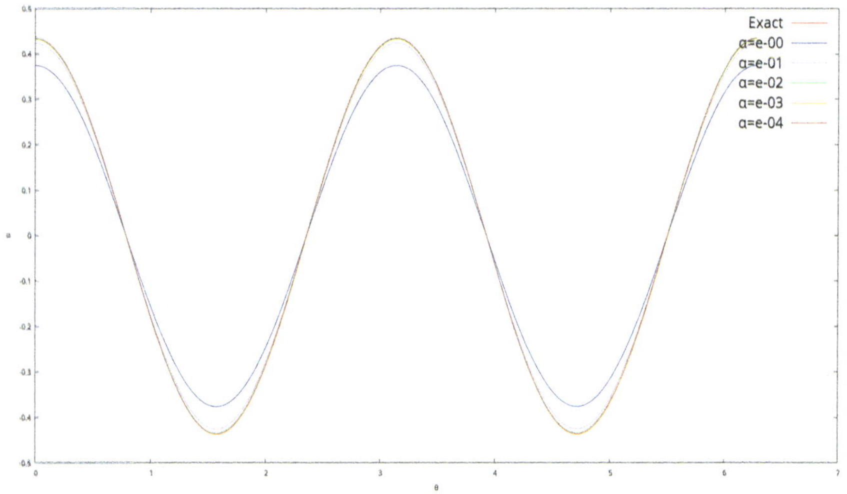

Fig. 2 Noisy solutions for different regularization values

Overall, the results emphasize that the regularized solution approach requires careful tuning of the regularization parameter and is computationally demanding due to the need for numerical integration and optimization of α.

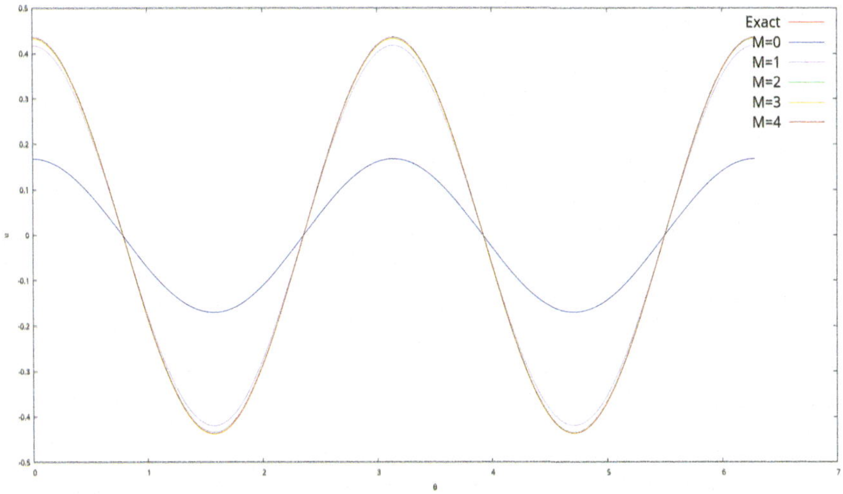

Fig. 3 HPM solutions for truncation values M

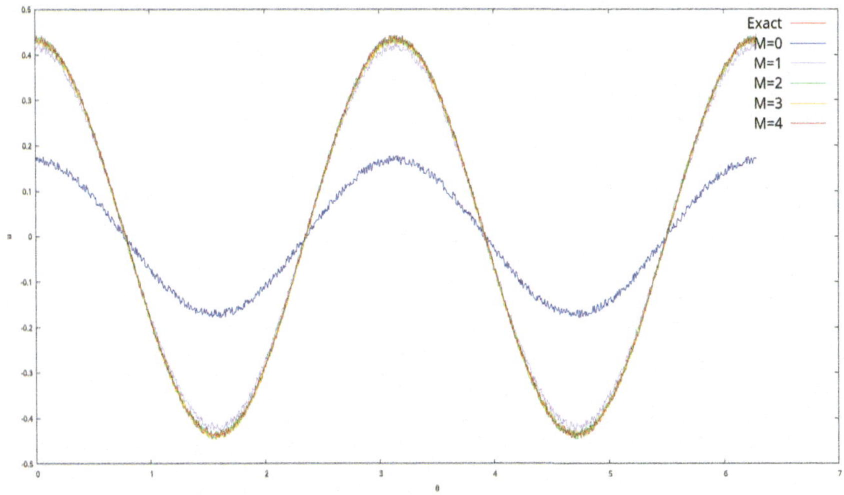

Fig. 4 Noisy HPM solutions for truncation values M

3.2 Homotopy Perturbation Method (HPM)

The homotopy perturbation method (HPM) demonstrated fast convergence, as seen in Fig. 3, where the solution was well-approximated using only three terms of the series expansion (6).

This efficiency makes HPM computationally appealing. When noise was added to the boundary condition g, the HPM solution remained robust, as shown in Fig. 4. The error observed in the solution was of the same order as the introduced noise (10^{-2}), indicating the regularizing nature of the method. Unlike the regularized approach, HPM does not depend on additional parameters like α; instead, its accuracy is influenced solely by the truncation order M of the series expansion (6).

Additionally, HPM is less computationally intensive, as it does not require numerical integration or parameter tuning, and its structure allows for efficient parallelization of calculations.

3.3 Comparison of the Two Methods

While both methods successfully approximate the exact solution, their performance varies under different conditions:

- The regularized solution approach achieves high accuracy in noiseless cases but is highly sensitive to noise and requires careful selection of the regularization parameter α.
- HPM demonstrates resilience to noise, achieving stable solutions without requiring parameter tuning. Its computational efficiency and simplicity in implementation further highlight its advantages.

4 Conclusion

In this study, we explored two analytical methods for solving the inverse Cauchy problem for Laplace's equation. The regularized solution approach is effective but sensitive to noise, requiring a well-chosen regularization parameter for stability. The homotopy perturbation method, on the other hand, is computationally efficient and inherently regularizing, making it more robust to noisy boundary conditions.

Overall, HPM is particularly suited for applications where computational speed and stability are critical, whereas the regularized approach may be preferred in cases where noise levels are minimal and high precision is required. These findings underline the complementary strengths of the two methods, suggesting their combined potential for tackling diverse inverse problems.

References

1. Jourhmane, M., Nachaoui, A.: An alternating method for an inverse cauchy problem. Numer. Algor. **21**(1), 247–260 (1999)
2. Pascual-Marqui, R.D.: Rev. Methods Solv. EEG Inv. Probl. **1**, 75–86 (1999)
3. Ellabib, A., Nachaoui, A.: On the numerical solution of a free boundary identification problem. Inv. Probl. Eng. **9**(3), 235–260 (2001)
4. Nachaoui, A.: An improved implementation of an iterative method in boundary identification problems. Numer. Algor. **33**(1–4), 381–398 (2003)
5. Fang, W., Lu, M.: A fast collocation method for an inverse boundary value problem. Int. J. Numer. Meth. Eng. **21**, 1563–1585 (2004)
6. Nachaoui, A.: Numerical linear algebra for reconstruction inverse problems. J. Comput. Appl. Math. **162**, 147–164 (2004)
7. Essaouini, M., Nachaoui, A., El Hajji, S.: Reconstruction of boundary data for a class of nonlinear inverse problems. J. Inv. Ill-Posed Probl. **12**(4), 369–385 (2004)
8. Regińska, T., Regiński, K.: Approximate solution of a cauchy problem for the helmholtz equation. Inverse Prob. **22**(3), 975–989 (2006)
9. Ellabib, A., Nachaoui, A.: An iterative approach to the solution of an inverse problem in linear elasticity. Math. Comput. Simul. **77**(2–3), 189–201 (2008)
10. Fu, C.L., Feng, X.L., Qian, Z.: The fourier regularization for solving the cauchy problem for the helmholtz equation. Appl. Numer. Math. **59**(10), 2625–2640 (2009)
11. Qian, A.L., Xiong, X.T., Wu, Y.-J.: On a quasi-reversibility regularization method for a cauchy problem of the helmholtz equation. J. Comput. Appl. Math. **233**(8), 1969–1979 (2010)
12. Berdawood, K.A., Nachaoui, A., Nachaoui, M., Aboud, F.: An effective relaxed alternating procedure for cauchy problem connected with helmholtz equation. Numer. Methods Part. Diff. Equ. **39**(3), 1888–1914 (2022)
13. Aboud, F., Nachaoui, A., Nachaoui, M.: On the approximation of a cauchy problem in a non-homogeneous medium. J. Phys.: Conf. Ser. **1743**(1), 012003 (2021)
14. Grunau, H.-C., Miyake, N., Okabe, S.: Positivity of solutions to the cauchy problem for linear and semilinear biharmonic heat equations. Adv. Nonlinear Anal. **10**(1), 353–370 (2021)
15. Berdawood, K., Nachaoui, A., Nachaoui, M.: An accelerated alternating iterative algorithm for data completion problems connected with helmholtz equationn. Stat. Optim. & Inf. Comput. **11**(1), 2–21 (2023). https://doi.org/10.1002/num.22793
16. Nachaoui, A., Nachaoui, M.: An hybrid finite element method for a quasi-variational inequality modeling a semiconductor. RAIRO Oper. Res. **57**(4), 2197–2207 (2023). https://doi.org/10.1051/ro/2022187
17. Nachaoui, A.: Iterative methods for inverse problems subject to the convection-diffusion equation. Adv. Math. Models & Appl. **8**, 327–338 (2023)
18. Nachaoui, A.: Cauchy's problem for the modified biharmonic equation : derivation and decoupled iterative methods. Springer Proceedings in Mathematics & Statistic, vol. 428. Springer, Cham (2025)
19. Ellabib, A., Nachaoui, A., Ousaadane, A.: A relaxed robin-robin algorithm for solving a cauchy problem in linear elasticity. J. Appl. Math. Comput. **71**, 1721–1741 (2025). https://doi.org/10.1007/s12190-024-02295-x
20. Hadri, A., Laghrib, A., Nachaoui, A., Nachaoui, M.: On the resolution of the non-smooth inverse cauchy problem by the primal-dual method. Numer. Algor. (2024). https://doi.org/10.1007/s11075-024-01950-5
21. Nachaoui, A., Sadiq, G.W.: Investigation of an inverse problem in heat propagation, To appear (2025)
22. Liu, J.C., Wei, T.: A quasi-reversibility regularization method for an inverse heat conduction problem without initial data. Appl. Math. Comput. **219**(23), 10866–10881 (2013)
23. Lattès, R., Lions, J.: The method of quasi-reversibility: applications to partial differential equations, Modern Analytic and Computational Methods in Science and Mathematics. American Elsevier Publishing Co., Inc, New York (1969)

24. Cheng, X., Gong, R., Han, W.: A coupled complex boundary method for the cauchy problem. Inverse Probl. Sci. Eng. **24**(9), 1510–1527 (2016)
25. Koshev, N., Yavich, N., Malovichko, M., Skidchenko, E., Fedorov, M.: Fem-based scalp-to-cortex eeg data mapping via the solution of the cauchy problem. J. Inv. Ill-Posed Probl. **28**(4), 517–532 (2020)
26. Malovichko, M., Koshev, N., Yavich, N., Razorenova, A., Fedorov, M.: Electroencephalographic source reconstruction by the finite-element approximation of the elliptic Cauchy problem. EEE Trans. Biomed. Eng. **68**(6), 1811–1819 (2020)
27. Nachaoui, A., Nachaoui, M., Tadumadze, T.: Meshless methods to noninvasively calculate neurocortical potentials from potentials measured at the scalp surface. Springer Proceedings in Mathematics & Statistics, vol. 428, pp. 99–117. Springer Cham (2023)
28. Liu, C.S., Wang, F.: A meshless method for solving the nonlinear inverse cauchy problem of elliptic type equation in a doubly-connected domain. Comput. Math. Appl. **76**, 1831–1852 (2018)
29. Nachaoui, A., Aboud, F., Nachaoui, M.: Acceleration of the kmf algorithm convergence to solve the cauchy problem for poisson's equation. Springer Proceedings in Mathematics & Statistics, vol. 372, pp. 43–57. Springer Cham (2021)
30. Nachaoui, A., Nachaoui, M., Chakib, A., Hilal, M.: Some novel numerical techniques for an inverse cauchy problem. J. Comput. Appl. Math. **381**(113030) (2021)
31. Ellabib, A., Nachaoui, A., Ousaadane, A.: Convergence study and regularizing property of a modified robin-robin method for the cauchy problem in linear elasticity. Inverse Prob. **38**(7), 075007 (2022)
32. Belytschko, T., Krongauz, Y., Organ, D., Fleming, M., Krysl, P.: Meshless methods: an overview and recent developments. Comput. Methods Appl. Mech. Engrg. **139**, 3–47 (1996)
33. Fairweather, G., Karageorghis, A.: The method of fundamental solutions for elliptic boundary value problems. Adv. Comput. Math. **9**, 69–95 (1998)
34. Shen, S., Atluri, S.N.: The Meshless Local Petrov-Galerkin (MLPG) Method (2002)
35. Jin, B., Zheng, Y.: A meshless method for some inverse problems associated with the helmholtz equation. Comput. Methods Appl. Mech. Eng. **195**(19–22), 2270–2288 (2006)
36. Wei, T., Hon, Y.C., Ling, L.: Method of fundamental solutions with regularization techniques for cauchy problems of elliptic operators. Eng. Anal. Boundary Elem. **31**(4), 373–385 (2007)
37. Liu, C.S., Atluri, S.N.: A highly accurate technique for interpolations using very high-order polynomials, and its applications to some ill-posed linear problems. Comput. Model. Eng. & Sci. **43**(1), 253–276 (2009)
38. Hua, Q., Gu, Y., Qu, W., Chen, W., Zhang, C.: A meshless generalized finite difference method for inverse cauchy problems associated with three-dimensional inhomogeneous helmholtz-type equations. Eng. Anal. Bound. Elem. **82**, 162–171 (2017)
39. Yue, X., Wang, F., Hua, Q., Qiu, X.: A novel space-time meshless method for nonhomogeneous convection-diffusion equations with variable coefficients. Appl. Math. Lett. **92**, 144–150 (2019)
40. Reddy, G.M.M., Nanda, P., Vynnycky, M., Cuminato, J.A.: An adaptive boundary algorithm for the reconstruction of boundary and initial data using the method of fundamental solutions for the inverse cauchy-stefan problem. Comput. Appl. Math. **40**(3) (2021)
41. Aboud, F., Jameel, I.T., Hasan, A.F., Mostafa, B.K., Nachaoui, A.: Polynomial approximation of an inverse cauchy problem for helmholtz type equations. Adv. Math. Models & Appl. **7**(3), 306–322 (2022)
42. Nachaoui, A., Aboud, F.: Solving geometric inverse problems with a polynomial based meshless method. Springer Proceedings in Mathematics & Statistics, vol. 428, pp. 119–136. Springer, Cham (2023)
43. Nachaoui, A., Rashid, S.M.: A mesh free wavelet method to solve the cauchy problem for the helmholtz equation. Springer Proceedings in Mathematics & Statistics, vol. 428, pp. 77–98. Springer Cham (2023)
44. Rashid, S.M., Nachaoui, A.: A haar wavelets-based direct reconstruction method for the cauchy problem of the poisson equation. Discrete Contin. Dyn. Syst. Ser. S **18**(1), 148–168 (2025)

45. Nachaoui, A., Sadiq, G.W.: Identification of a space dependent force in a hyperbolic equation using a polynomial expansion. Springer Proceedings in Mathematics & Statistics, vol. 428. Springer, Cham (2025)
46. Liu, C.-S.: Solving the inverse problems of laplace equation to determine the robin coefficient/cracks' position inside a disk. Comput. Model. Eng. & Sci. **40**(1), 1–28 (2009)
47. Nachaoui, A., Salih, H.W.: An analytical solution for the nonlinear inverse cauchy problem. Adv. Math. Models Appl. **6**(3), 191–206 (2021)
48. Nachaoui, A., Rasheed, S.M., Sadiq, G.W.: A semi-analytical solution for inverse cauchy problems. Springer Proceedings in Mathematics & Statistics, vol. 428. Springer, Cham (2025)
49. He, J.-H.: Homotopy perturbation technique. Comput. Methods Appl. Mech. Eng. **178**(3), 257–262 (1999)

Simultaneous Recovery of Corroded Boundaries and Admittance Using the Kohn–Vogelius Method

Moustapha Essahraoui, Elmehdi Cherrat, Lekbir Afraites, and Julius Fergy Tiongson Rabago

Abstract We address the problem of identifying an unknown portion Γ of the boundary of a d-dimensional ($d \in \{1, 2\}$) domain Ω and its associated Robin admittance coefficient, using two sets of boundary Cauchy data (f, g)-representing boundary temperature and heat flux–measured on the accessible portion Σ of the boundary. Identifiability results [9, 47] indicate that a single measurement on Σ is insufficient to uniquely determine both Γ and α, but two independent inputs yielding distinct solutions ensure the uniqueness of the pair Γ and α. In this paper, we propose a cost function based on the energy-gap of two auxiliary problems. We derive the variational derivatives of this objective functional with respect to both the Robin boundary Γ and the admittance coefficient α. These derivatives are utilized to develop a nonlinear gradient-based iterative scheme for the simultaneous numerical reconstruction of Γ and α. Numerical experiments are presented to demonstrate the effectiveness and practicality of the proposed method.

Keywords Geometric inverse problem · Robin boundary condition · Shape optimization · Shape derivatives · Simultaneous recovery

1 Introduction

We consider a classical inverse geometry problem arising in non-destructive testing and evaluation, specifically in the detection of an internally corroded boundary. Our objective is to determine a doubly connected domain Ω in $\mathbb{R}^d$, $d \in \{2, 3\}$, externally bounded by the accessible boundary Σ and internally by the unknown boundary Γ, along with the coefficient α that appears in the boundary condition of the PDE

M. Essahraoui · E. Cherrat · L. Afraites (✉)
Mathematics and Interactions Teams (EMI), Faculty of Sciences and Techniques, Sultan Moulay Slimane University, Beni Mellal, Morocco
e-mail: l.afraites@usms.ma

J. F. T. Rabago
Faculty of Mathematics and Physics, Institute of Science and Engineering, Kanazawa University, Kanazawa, Japan

A. Laghrib et al. (eds.), *Computational Methods for Inverse Problems and Applications*, Springer Proceedings in Mathematics & Statistics 498,
https://doi.org/10.1007/978-3-031-89498-5_6

model. This is done using two pairs of Cauchy data, $\{(f_k, g_k)\}, k = 1, 2$, on Σ for a harmonic function u defined in Ω. On the unknown boundary Γ, the function u is assumed to satisfy a homogeneous Robin boundary condition. Consequently, given Σ and $\{(f_k, g_k)\}, k = 1, 2$, we examine the overdetermined boundary value problem:

$$-\Delta u_k = 0 \quad \text{in } \Omega, \quad u_k = f_k, \quad \frac{\partial u_k}{\partial n} = g_k \quad \text{on } \Sigma, \quad \frac{\partial u_k}{\partial n} + \alpha u_k = 0 \quad \text{on } \Gamma, \quad (1)$$

where $\alpha \in L^\infty(\mathbb{R}^d)$ such that $\alpha(x) \geqslant \alpha_0 > 0$ for all $x \in \Gamma$, and α_0 is a known constant. The vector n denotes the outward unit normal to $\partial\Omega$.

In practical applications, the goal is to identify the inaccessible boundary using electrostatic measurements or thermal imaging techniques on the externally accessible part Σ. For instance, u can be interpreted as the electrostatic potential in a specimen Ω, where only the boundary portion Σ is accessible for measurements. In other words, (1) can be interpreted as the determination of the shape of the inaccessible boundary portion Γ based on two sets of imposed voltages $u_k|_\Sigma$ and the corresponding measured resulting currents $\partial_n u_k|_\Sigma$ on Σ. Thus, in the inverse problem setting (see, e.g., [26, 41]), we are interested in solving the following:

Problem 1 Determine the unknown boundary Γ, the Robin coefficient α, and the electrostatic potentials u_k, for $k = 1, 2$, that satisfy the Cauchy data problem:

$$-\Delta u_k = 0 \quad \text{in } \Omega, \quad u_k = f_k, \quad \frac{\partial u_k}{\partial n} = g_k \quad \text{on } \Sigma, \quad \frac{\partial u_k}{\partial n} + \alpha u_k = 0 \quad \text{on } \Gamma.$$

Extensive theoretical and numerical investigations have been conducted on the inverse problem of determining α from a single measurement of u on a sub-boundary of Σ, assuming that Γ is known. Notable works in this area include [12, 13, 27, 33, 34, 37, 38, 40, 43], among others. In contrast, numerical studies on recovering Γ from measurements on part or all of Σ, assuming α is known, are discussed in [5, 6, 21, 28, 29, 32, 48]. Additionally, the more general problem of simultaneous reconstruction of both Γ and α has been addressed in [9, 47, 52]. Analytical and numerical recovery approaches, based on the equivalent boundary integral equation formulation of the problem, are presented in [17–19]. Recently, a numerical shape optimization approach for the simultaneous recovery of Γ and α, using two Cauchy pairs and leveraging shape derivatives, was introduced in [25]. Motivated by this study, we propose an energy-gap cost functional for the simultaneous recovery of the unknown inaccessible interior boundary Γ and the Robin coefficient α, as an alternative to the least-squares boundary tracking functional used in [25].

Let us briefly explain why two sets of Cauchy data are required in Problem 1, which relates to the identifiability issue. Cakoni and Kress [17] showed that, for constant α, a single Cauchy pair (f, g) on Σ may correspond to infinitely many domains Ω. They provided counterexamples demonstrating that one pair is insufficient to uniquely determine both Γ and α. On the other hand, Bacchelli [9] proved that two linearly independent Cauchy pairs ensure the unique identification of Γ and α, provided one input is positive. Similarly, Pagani and Pierotti [47] established uniqueness using two measurements. On a different note, for stability results with two independent inputs, we refer readers to to Sincich [52].

Problem setting. Let us now present our problem in a mathematically precise manner. We let $D \subset \mathbb{R}^d$, $d \in \{2, 3\}$, be a $C^{2,1}$ bounded open set and for a fixed real number $\delta > 0$, define the admissible set of unknown inclusions $\mathcal{U}$ as follows:

$$\mathcal{U} := \left\{\omega \Subset D \mid \omega \in C^{2,1},\ d(x, \partial D) > \delta\ \forall x \in \omega,\ \text{ and } D \setminus \overline{\omega} \text{ is connected}\right\}.$$

We denote $\Omega := D \setminus \overline{\omega}$, $\Sigma := \partial D$, and $\Gamma := \partial\omega$, and assume (only for technical purposes; see [6]) that $f \in H^{5/2}(\Sigma)$, $f \not\equiv 0$. Moreover, we let $g \in H^{3/2}(\Sigma)$ be an admissible boundary measurement corresponding to f. Furthermore, we restrict of investigation to the following admissible set of Robin parameters:

$$\mathcal{A} := \left\{\alpha \in L^\infty(\Gamma) \mid \alpha_0 \leqslant \alpha \leqslant \alpha_1 \text{ for a.e. } x \in \Gamma\right\}.$$

In this study, we tacitly assumed that we can find $(\omega^*, \alpha^*) \in \mathcal{U} \times \mathcal{A}$ such that Problem 1 has a solution. That is, we assume that the surface measurement g (or f, if g is given instead) is obtained without error. Therefore, we precisely consider the following inverse geometry problem here:

Problem 2 Find $(\omega, \alpha) \in \mathcal{U} \times \mathcal{A}$ and u such that (1) is satisfied.

To address Problem 2 we define the functional J as

$$J(\omega, \alpha) := \sum_{k=1}^{2} \frac{1}{2} \int_\Omega |\nabla(u_{D,k} - u_{N,k})|^2\, dx + \frac{1}{2} \int_\Gamma \alpha |u_{D,k} - u_{N,k}|^2\, ds, \tag{2}$$

and reformulate the overdetermined problem (1) as the minimization problem:

Problem 3 Minimize $J(\omega, \alpha)$ subject to (3) and (4), where $u_{N,k} := u_{N,k}(\omega, \alpha)$ and $u_{D,k} := u_{D,k}(\omega, \alpha)$ respectively satisfy

$$-\Delta u_{N,k} = 0 \text{ in } \Omega, \qquad \partial_n u_{N,k} = g_k \text{ on } \Sigma, \qquad \partial_n u_{N,k} + \alpha u_{N,k} = 0 \text{ on } \Gamma, \tag{3}$$

$$-\Delta u_{D,k} = 0 \text{ in } \Omega, \qquad u_{D,k} = f_k \text{ on } \Sigma, \qquad \partial_n u_{D,k} + \alpha u_{D,k} = 0 \text{ on } \Gamma. \tag{4}$$

Problem 3 is equivalent to Problem 1 when, for all $k = 1, 2$, $u_k = f_k$ and $\partial_n u_k = g_k$ on Σ (cf. [6, Remark 2.6]).

The functional J is known as the Kohn–Vogelius functional [42]. It has been applied in various contexts, including shape detection in electrical impedance tomography [23], transmission problems [2, 3], inverse Robin boundary problem [6], inverse geometry problems in Stokes flow [11], and geometric inverse source problems [4]. See also [1, 10, 24] for the application of the method to solving the exterior Bernoulli free boundary problem. Notably, this volume-based cost functional often yields more accurate solutions than boundary-integral-based alternatives.

Paper organization. The remainder of the paper is organized as follows. Section 2 discusses the sensitivity of J with respect to the unknown inclusion ω and the unknown Robin coefficient α. Section 3 presents a numerical algorithm based on the shape derivative of J with respect to ω and the Fréchet derivative of J with respect to the parameter α. The proposed numerical method is then tested through various numerical examples.

2 Sensitivity of J Respect to ω and α

In this section, we present the derivative of $J(\omega, \alpha)$ with respect to ω, followed by its derivative with respect to α. For notational convenience, when one argument is fixed, we omit the functional's dependence on that argument.

2.1 Shape Derivative of J with Respect to ω

Let us consider Problem 3 and suppose that $\alpha \in \mathcal{A}$ is fixed. For $\varphi, \psi \in H^1(\Omega)$, we define

$$a(\alpha; \varphi, \psi) = \int_{\Omega} \nabla\varphi \cdot \nabla\psi \, dx + \int_{\Gamma} \alpha\varphi\psi \, ds.$$

The variational formulations of (3) are (4) are as follows:

$$\text{Find } u_D \in H^1(\Omega), u_D|_{\Sigma} = f, \text{ such that } a(\alpha; u_D, \psi) = 0, \forall\psi \in V(\Omega); \tag{5}$$

$$\text{Find } u_N \in H^1(\Omega) \text{ such that } a(\alpha; u_N, \psi) = \langle g, \psi\rangle_{\Sigma}, \forall\psi \in H^1(\Omega), \tag{6}$$

The weak forms (5) and (6) are known to be well-posed if $(f, g) \in H^{1/2}(\Sigma) \times H^{-1/2}(\Sigma)$ and Ω is Lipschitz regular, as established by the Lax-Milgram lemma (see, e.g., [6, p. 46]). Meanwhile, the existence of a minimizer for J can be shown under an appropriate topology, provided that the admissible domain satisfies the a *uniform cone property* [35, Theorem 2.4.7, p. 56]. While we do not prove this claim here, the arguments used in [5, 49] can be followed to verify it.

To numerically solve Problem 3, we will use a shape-gradient-based technique combined with the finite element method (FEM). To implement this, we require the expression for the shape derivative of J, which has already been derived in [6, Prop. 3.4]. Therefore, we direct the reader to [6] for the detailed characterization of the shape gradient.

Proposition 1 (Shape gradient of J) *Let D_δ be an open set with a C^∞ boundary, such that $\{x \in D \mid d(x, \partial D) > \delta/2\} \subset D_\delta \subset \{x \in D \mid d(x, \partial D) > \delta/3\}$. Define Θ as the collection of $C^{2,1}(\mathbb{R}^d)^d$ vector fields with compact support in $\overline{D}_\delta$. Assume that $\Omega \in C^{2,1}$ is an admissible domain, i.e., $\Omega = D \setminus \overline{\omega}$ where $\omega \in \mathcal{U}$ and $\theta \in \Theta$. Then,* *the map* $t \mapsto J(\omega_t)$*, is* C^1 *in a neighborhood of* 0*, and its shape derivative at* 0 *is given by*

$$dJ(\omega)[\theta] = \int_\Gamma \left[-F(u_N, w) + |\nabla w|^2 + \frac{\alpha}{2}\left(2w\partial_n w + \kappa w^2\right)\right] \theta_n \, ds =: \langle G, \theta_n \rangle_\Gamma,$$

where $w = u_D - u_N$*, and* $F(v, q) = -\nabla_\Gamma v \cdot \nabla_\Gamma q - \alpha\,(\partial_n v + \kappa v)\, q - \partial_n \alpha v q$*, for* $v, q \in H^2(\Omega)$*. Here,* ∇_Γ *denotes the tangential gradient operator while* κ *represents the mean curvature of the free boundary* $\Gamma = \partial\omega$*.*

An important feature of the Kohn–Vogelius functional J is that its gradient does not depend on the derivatives of the states. This is a characteristic commonly found in energy-type cost functionals, allowing the gradient to be computed numerically without the need for adjoint variables. We highlight that this approach differs from the one presented in [25], which requires the use of adjoint variables. From a numerical standpoint, our method only needs for a regularization functional for α, in contrast to [25], as the proposed objective functional inherently includes a smoothing effect. This enhances stability and provides a more accurate approximation of the minimizer, offering a solution to the inverse problem of simultaneously recovering α and Γ.

2.2 Derivative of J with Respect to α

Let Ω be a given admissible domain. For simplicity, we denote the derivative of u_i with respect to α by $\widetilde{u}_i = \frac{\partial u_i}{\partial \alpha}$. Hereinafter, without further notice, we assume $i \in D, N$, indicating the solutions to Eqs. (4) and (3). Given $\alpha, \rho \in \mathcal{A}$, there exists a real number $\varepsilon = \varepsilon(\alpha, \rho) > 0$ such that $\alpha^\varepsilon = \alpha + \varepsilon\rho \in \mathcal{A}$. Consequently, α^ε is a perturbation of α in the direction ρ.

Proposition 2 ([13]) *Let* Ω *be a given admissible domain and* $\alpha, \rho \in \mathcal{A}$*. There exist* $\widetilde{u}_i, O(\varepsilon^2) \in H^1(\Omega)$ *such that*

$$u_i^\varepsilon = u_i + \varepsilon \widetilde{u}_i + O(\varepsilon^2) \quad \text{and} \quad \lim_{\varepsilon \searrow 0} O(\varepsilon^2) = 0, \tag{7}$$

where $\widetilde{u}_i$ *solves the variational equation*

$$a(\widetilde{u}_i, \psi) = -\int_\Gamma \rho u_i \psi \, ds, \tag{8}$$

for all $\psi \in H^1(\Omega)$ *when* $i = N$ *and for all* $\psi \in V(\Omega)$ *when* $i = D$*. Additionally,* J *is Gateaux differentiable at* α *in the* ρ *direction, and its derivative is given by*

$$dJ(\alpha)\rho = \frac{1}{2}\int_\Gamma \rho(u_D^2 - u_N^2)\, ds.$$

Proof Let Ω be a given admissible domain and $\alpha, \rho \in \mathcal{A}$. The differentiability of the states and the cost functional with respect to α can be shown by standard arguments, so we omit it. Let us first establish the structure of the derivatives of the states with respect to α. To do this, we let $\varepsilon = \varepsilon(\alpha, \rho) > 0$ be such that $\alpha^\varepsilon = \alpha + \varepsilon\rho \in \mathcal{A}$ and consider the following perturbed state problems:

$$a(\alpha^\varepsilon; u_N^\varepsilon, \psi) = \int_\Sigma g\psi \, ds, \quad \forall \psi \in H^1(\Omega), \tag{9}$$

$$a(\alpha^\varepsilon; u_D^\varepsilon, \psi) = 0, \quad \forall \psi \in V(\Omega). \tag{10}$$

Note that g represents an observation on the boundary Σ. Since the perturbation of α does not affect g, we have $g^\varepsilon = g$ on Σ.

Subtracting (3) from (9), we obtain

$$a(\alpha; u_N^\varepsilon - u_N, \psi) = -\int_\Gamma \varepsilon\rho u_N \psi \, ds, \qquad \forall \psi \in H^1(\Omega). \tag{11}$$

For any $\varepsilon > 0$, Eq. (11) has a unique weak solution $u_N^\varepsilon - u_N \in H^1(\Omega)$. By substituting $\psi = u_N^\varepsilon - u_N$ into the above equation, we deduce that $\left\| u_N^\varepsilon - u_N \right\|_{H^1(\Omega)} = O(\varepsilon)$, which suffices for computing $dJ(\alpha)\rho$. Additionally, the strong convergence of $\left\{ \frac{1}{\varepsilon}(u_N^\varepsilon - u_N) \right\}$ in $H^1(\Omega)$ can be established, though we omit the proof here.

To proceed, we divide Eq. (11) by ε and take the limit as $\varepsilon \to 0$. Utilizing (7), we deduce that $\widetilde{u}_N \in H^1(\Omega)$ satisfies the weak formulation

$$a(\alpha; \widetilde{u}_N, \psi) = -\int_\Gamma \rho u_N \psi \, ds, \quad \forall \psi \in H^1(\Omega).$$

Similarly, by applying the same approach, we find that $\widetilde{u}_D \in V(\Omega)$ satisfies

$$a(\alpha; \widetilde{u}_D, \psi) = -\int_\Gamma \rho u_D \psi \, ds, \quad \forall \psi \in V(\Omega).$$

Let us denote $w = u_D - u_N$ and $w^\varepsilon = u_D^\varepsilon - u_N^\varepsilon$. Now, the derivative of $J(\alpha)$ with respect to α in the direction ρ is given by

$$dJ(\alpha)\rho = \lim_{\varepsilon \searrow 0} \frac{J(\alpha + \varepsilon\rho) - J(\alpha)}{\varepsilon} = \lim_{\varepsilon \searrow 0} \frac{1}{2\varepsilon} \left[a(\alpha^\varepsilon; w^\varepsilon, w^\varepsilon) - a(\alpha; w, w) \right].$$

Observe that the expression on the right contains the difference $(u_i^\varepsilon - u_i)/\varepsilon = \widetilde{u}_i + O(\varepsilon)$. Hence, evaluating the limit leads to the expression

$$\begin{aligned} dJ(\alpha)\rho &= a(\alpha; \widetilde{w}, w) + \frac{1}{2} \int_\Gamma \rho |w|^2 \, ds \\ &= a(\alpha; \widetilde{u}_D, u_D) - a(\alpha; \widetilde{u}_D, u_N) - a(\alpha; \widetilde{u}_N, w) + \frac{1}{2} \int_\Gamma \rho |w|^2 \, ds. \end{aligned}$$

We eliminate $\widetilde{w} \in H^1(\Omega)$ appearing on the above expression. To do this, first, we consider (8) for $i = N$ and choose $\psi = w = u_D - u_N \in H^1(\Omega)$ to obtain

$$-a(\alpha; \widetilde{u}_N, w) = \int_\Gamma \rho u_N (u_D - u_N)\, ds.$$

Next, we consider (5) and take $\psi = \widetilde{u}_D \in V(\Omega) \subset H^1(\Omega)$ to get

$$a(\alpha; u_D, \widetilde{u}_D) = 0.$$

Finally, we consider (6) and choose $\psi = \widetilde{u}_D \in V(\Omega) \subset H^1(\Omega)$ (i.e., $\widetilde{u}_D|_\Sigma = 0$) to obtain

$$a(\alpha; \widetilde{u}_D, u_N) = a(\alpha; u_N, \widetilde{u}_D) = \langle g, \widetilde{u}_D \rangle_\Sigma = 0.$$

Thus, we arrive at the following simplification (i.e., eliminating $\widetilde{w}$)

$$\begin{aligned} dJ(\alpha)\rho &= \int_\Gamma (\rho u_N u_D - \rho u_N^2)\, ds + \frac{1}{2}\int_\Gamma \rho(u_D^2 - 2u_D u_N + u_N^2)\, ds \\ &= \frac{1}{2}\int_\Gamma \rho(u_D^2 - u_N^2)\, ds. \end{aligned}$$

That is, $dJ(\alpha) = \frac{1}{2}(u_D^2 - u_N^2)$. This proved the proposition.

3 Numerical Algorithm and Examples

3.1 Numerical Algorithm

Our numerical approach is built upon the gradient informations derived in the previous section, incorporating techniques from earlier works (see, e.g., [5, 6]) that implement a Sobolev-gradient-based algorithm in a finite element setting. Below, we provide the key details of our numerical method.

Choice of domain deformation vector. The choice $\theta = V_n n = -Gn$, where $G \in L^2(\Gamma)$ and $G \not\equiv 0$, provides a descent direction for the cost function J. However, if the data (f, g) lack sufficient smoothness, the shape gradient's surface expression given in Proposition 1 may not exist, and the descent vector θ could have poor regularity. Specifically, the $L^2(\Gamma)$ regularity of the shape gradient G is insufficient to achieve a stable approximation of the unknown boundary. To address this issue, we employ the Riesz representation of the shape gradient, a well-established technique in the literature (see, e.g., [7, 22]). To ensure a smooth descent direction for J, we determine a vector $\theta \in H^1_{\Sigma,0}(\Omega)^d$ that solves the variational equation:

$$\int_{\Omega} (\nabla\theta : \nabla\varphi + \theta \cdot \varphi)\, dx = -\int_{\Gamma} Gn \cdot \varphi\, ds, \quad \forall\varphi \in H^1_{\Sigma,0}(\Omega)^d.$$

This yields a *Sobolev gradient* [8, 46] representation θ of $-Gn$ over Ω.

Step-size computation and stopping condition. The kth step size t^k is computed at each iteration using a backtracking line search with the formula

$$t^k = \mu \frac{J(\omega^k, \alpha^k)}{\|\theta^k\|^2_{H^1(D\backslash\overline{\omega}^k)^d}},$$

where $\mu > 0$ is a scaling factor; see [50, p. 281]. The value of μ is adjusted to prevent the formation of inverted triangles in the mesh after the update. Meanwhile, ε^k is determined using the standard Armijo line search procedure. The algorithm terminates after a finite number of iterations. For precise measurements, we set the maximum number of iterations to 500, and for noisy data, we reduced it to 200.

Choice of regularization functional. A common regularization approach involves adding specific terms to the minimization process, particularly for addressing ill-posed systems [25, 51]. In our experiments, we employ a single regularization term based on the PDE formulation, alongside the previously mentioned extension-regularization technique. This approach is independent of the parametrization of (Γ, α) or the choice of numerical solver, simplifying the process by reducing the need for tuning multiple parameters. One possible regularization functional is given by:

$$\frac{\tau_1}{2} P(\Gamma) + \frac{\tau_2}{2} R(\alpha) = \frac{\tau_1}{2} \int_{\Gamma} 1\, ds + \frac{\tau_2}{2} \int_{\Gamma} \alpha^2\, ds, \tag{12}$$

where $\tau_1, \tau_2 > 0$ are constants. However, we omit the perimeter term and rely solely on Tikhonov regularization R for α. As demonstrated in our examples, this choice is sufficient for achieving reasonable reconstruction, even in the presence of noisy data.

Choice of regularization parameter. When reconstructing noisy data, choosing the regularization parameter $\tau = \tau_2$ in (12) is crucial. This parameter is typically selected using the discrepancy principle, which requires accurate knowledge of the noise level. However, in many cases, the noise level is either unknown or unreliable, and an incorrect estimate can significantly degrade reconstruction accuracy. To overcome this, heuristic rules for parameter selection are needed, especially when noise level information is unavailable or unreliable. We propose a heuristic rule based on the balancing principle [16], which selects $\tau > 0$ such that

$$(\beta - 1) J(\omega, \alpha) - \frac{\tau}{2} R(\mu) := (\beta - 1)\frac{1}{2} a(\alpha; w, w) - \frac{\tau}{2} \|\alpha\|^2_{L^2(\Gamma)^2} = 0, \tag{13}$$

for a constant $\beta > 1$. This balances the data-fitting term $J(\omega, \alpha)$ with the penalty term $R(\alpha)$, where β controls the trade-off. This rule does not rely on noise level information and has been successfully applied to both linear and nonlinear inverse problems [14–16, 20, 36], and recently in [44, 45].

Algorithm. The kth pair approximation (ω^k, α^k) of the exact solution pair (ω^*, α^*) is computed as follows:

Algorithm 1 Simultaneous recovery of Γ and α

1: **Initialization:** Fix $D \subset \mathbb{R}^d$, τ, $\beta > 0$, and choose an initial shape ω^0 and Robin coefficient α^0.
2: **for** $k = 0, 1, 2, \ldots$ **do**
3: Solve the two-state problems on the $D \setminus \overline{\omega}^k$ with Robin coefficient α^k.
4: Choose $t^k > 0$ and compute the vector θ^k in $D \setminus \overline{\omega}^k$.
5: Update the inclusion: $\omega^{k+1} = (\mathrm{id} + t^k \theta^k)\omega^k$.
6: Compute τ via the balancing principle (13).
7: Choose $\varepsilon^k > 0$ and update the Robin coefficient: $\alpha^{k+1} = \alpha^k - \varepsilon^k dJ(\alpha) + \varepsilon^k \tau \alpha^k$.
8: **end for**

3.2 Numerical Examples

For the numerical examples, we make the following broad assumptions: The specimen under examination has a circular shape with a unit radius, centered at the origin; i.e., $D = B(0, 1)$. For $(x_1, x_2) \in \Gamma^*$, we consider three cases for the exact Robin coefficient:

Case A1: $\alpha_1^* = \exp(x_1 x_2)$;
Case A2: $\alpha_2^* = 1 + 0.5 x_1 x_2$;
Case A3: $\alpha_3^* = 1 + 0.5 \sin(\pi x_1) \sin(\pi x_2)$.

The data is generated synthetically. Specifically, we impose the Dirichlet boundary conditions $f_1(t) = \cos(t)$ and $f_2(t) = \sin(t)$, $t \in [0, 2\pi]$, and set $g_1 = \partial_n u_1^*|_\Sigma$ and $g_2 = \partial_n u_1^*|_\Sigma$, respectively.

To prevent "inverse crimes" (see [39, p. 154]), we use different numerical methods for generating synthetic data and performing the inversion. For the forward problem, the domain is discretized using 300 nodal points on both the exterior and interior boundaries, with P_2 finite element basis functions implemented in FreeFem++ [31]. In contrast, the inversion process uses a uniform mesh with 150 nodes on each boundary, a mesh size of $h = 0.05$, and P_1 finite elements to solve all variational problems.

We will evaluate the proposed identification procedure by considering the following geometries for the unknown boundary Γ:

$$\textbf{Case B1:} \quad \Gamma_1^* = \left\{ \begin{pmatrix} 0.1 + 0.7\cos t \\ 0.2 + 0.5\sin 2t \end{pmatrix}, \forall t \in [0, 2\pi) \right\};$$

$$\textbf{Case B2:} \quad \Gamma_2^* = \left\{ \begin{pmatrix} 0.6\cos t \\ 0.5\sin t(1.8 + \cos 2t) \end{pmatrix}, \forall t \in [0, 2\pi) \right\};$$

$$\textbf{Case B3:} \quad \Gamma_3^* = \left\{ \begin{pmatrix} -0.25 + \dfrac{0.6 + 0.54\cos t + 0.06\sin 2t}{1 + 0.75\cos t}\cos t \\ 0.05 + \dfrac{0.6 + 0.54\cos t + 0.06\sin 2t}{1 + 0.75\cos t}\sin t \end{pmatrix}, \forall t \in [0, 2\pi) \right\}.$$

In all cases, the algorithm starts with $\alpha^0 = 1$ and an initial guess of $\Gamma^0 = C(\mathbf{0}, r_0)$, where $r_0 \in \{0.5, 0.55, 0.6\}$, depending on the case. It is terminated after 500 iterations in all cases. Although the algorithm could run longer and the stopping criterion could be improved, this simple approach already provides satisfactory results based on our experience. We also test the algorithm on noisy data by perturbing the exact measurements with Gaussian noise of mean zero and standard deviation 0.5, where the noise level is controlled by the parameter δ (noise weight). Due to the severely ill-posed nature of the problem, we limit our analysis to a noise level of $\delta = 0.005$.

Table 1 Reconstructed of shapes

Case	A1	A2	A3
B1	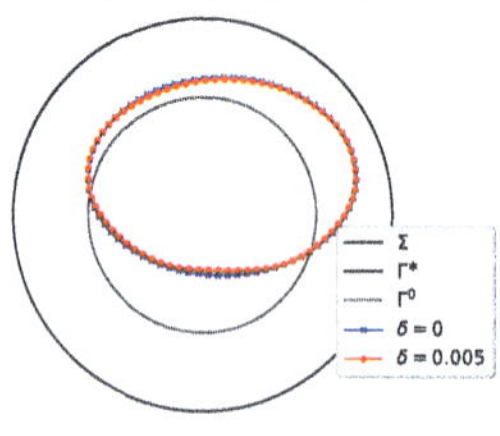	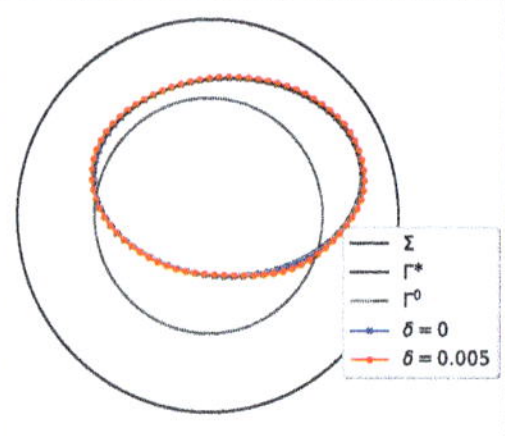	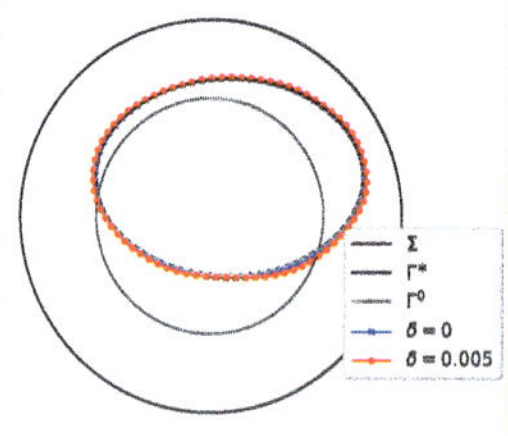
B2	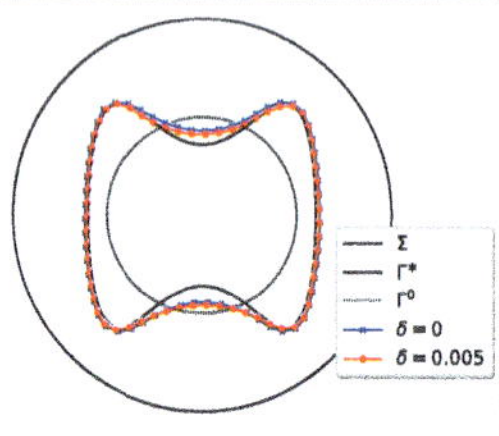	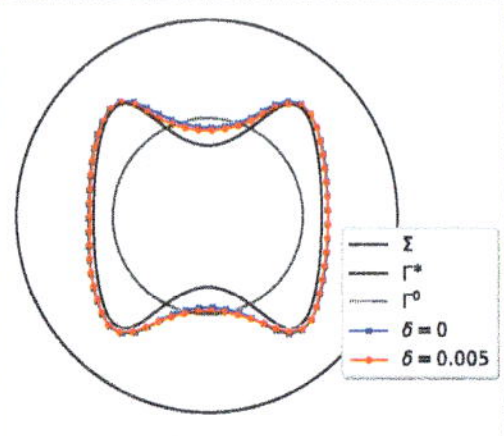	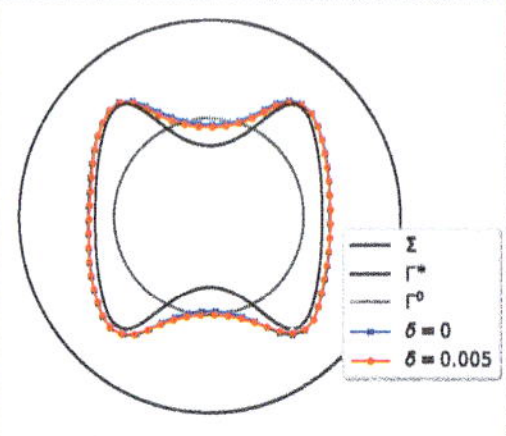
B3	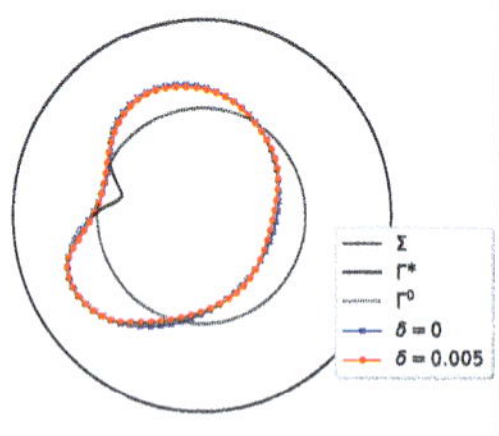	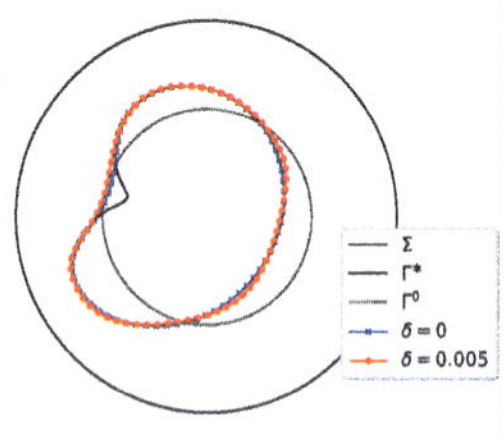	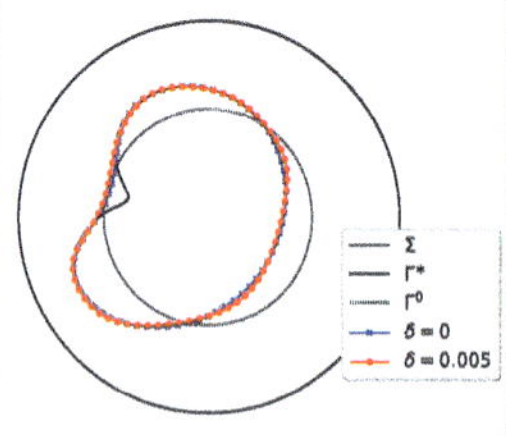

Table 2 3D view of the reconstructed Robin coefficient, rotated by 30°

Case	A1	A2	A3
B1	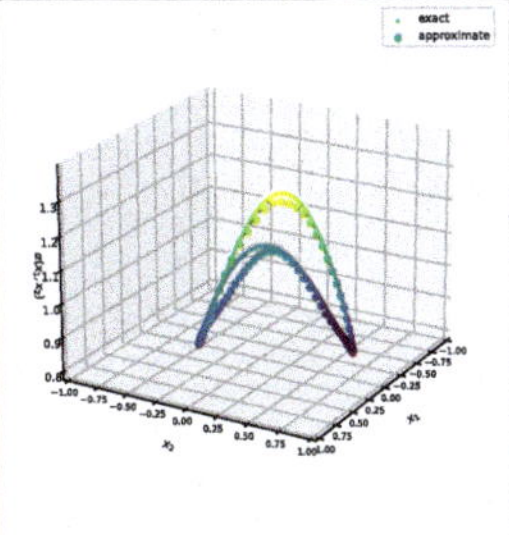	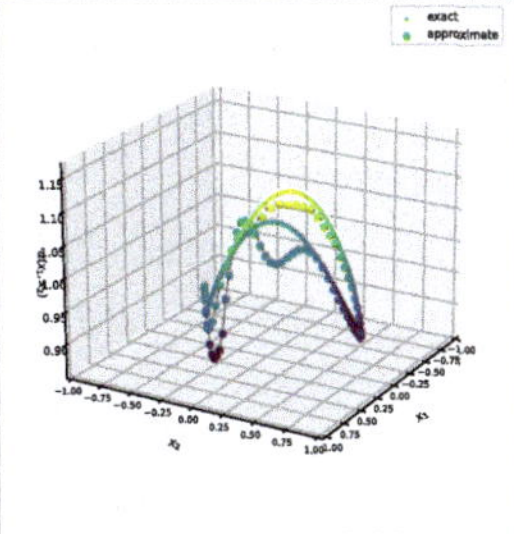	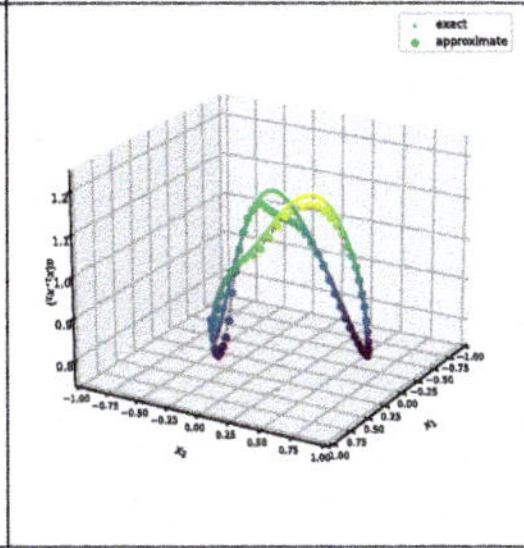
B2	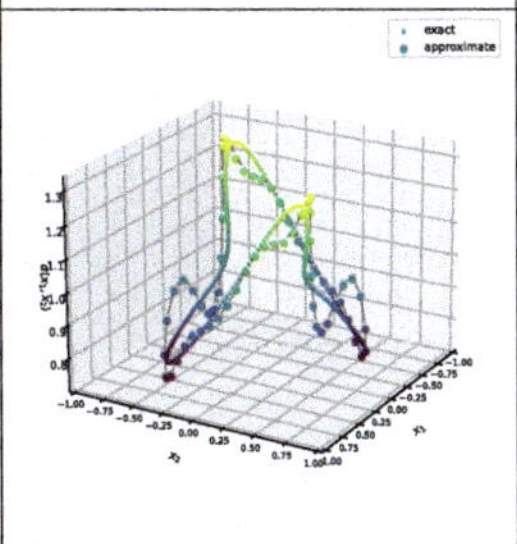	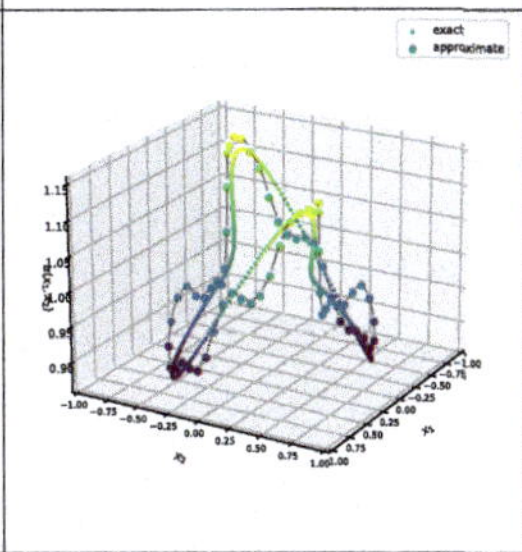	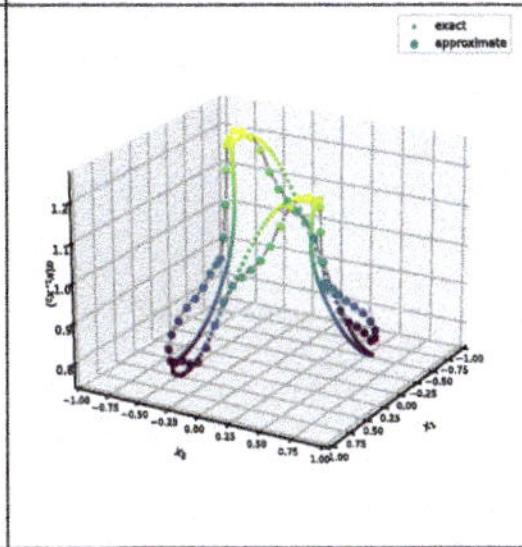
B3	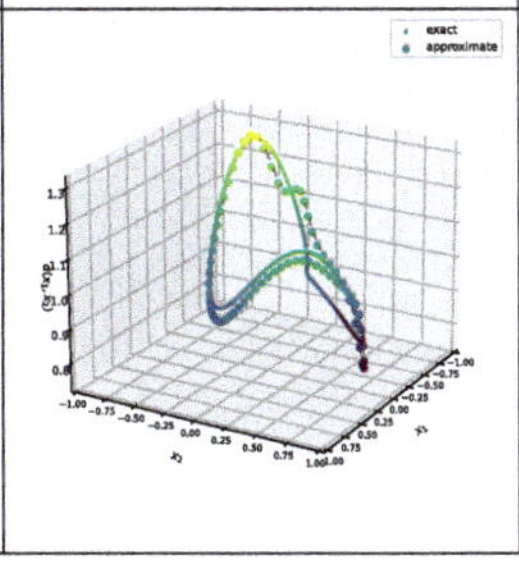	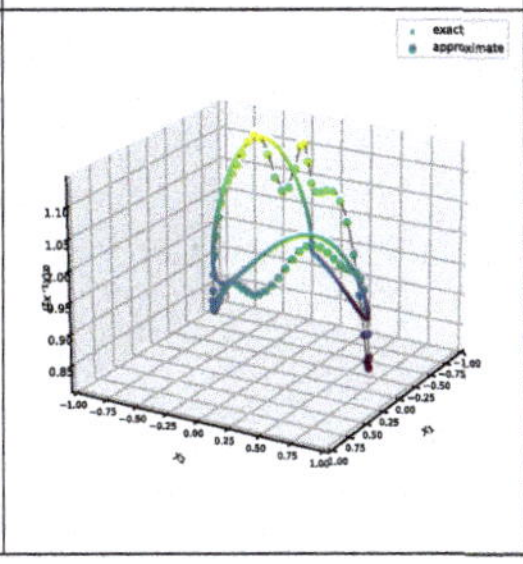	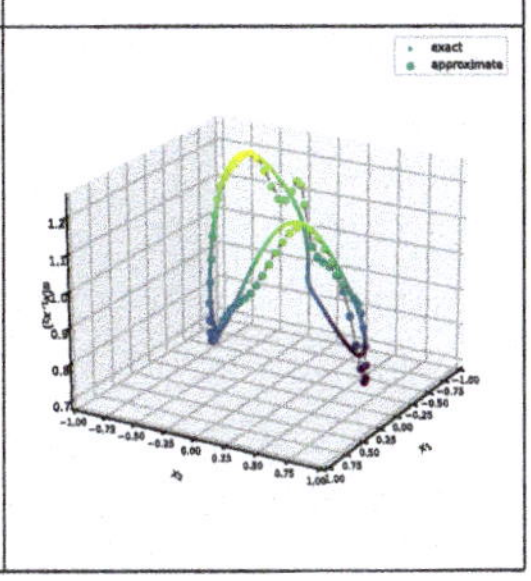

3.3 *Numerical Results and Discussion*

In this subsection, we discuss the results of our numerical experiments. Table 1 shows the reconstructed shapes for Cases A1–A3 combined with Cases B1–B3. Each row corresponds to a specific case, comparing the shapes obtained from exact and noisy data against the true inclusion. The outermost thick black line indicates the object's surface, while the inner black line shows the exact geometry of the unknown boundary. The black dashed lines represent the initial guesses. The blue line with "×" markers and the red dashed line with "◯" markers depict the reconstructions under exact and noisy ($\delta = 0.005$) measurements, respectively. We observe that the method achieves reasonable reconstructions for test problems, especially without noise; see

Table 3 3D view of the reconstructed Robin coefficient, rotated by 135°

Case	A1	A2	A3
B1	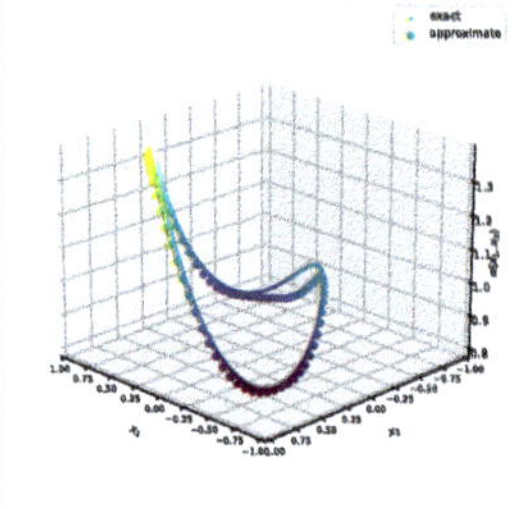	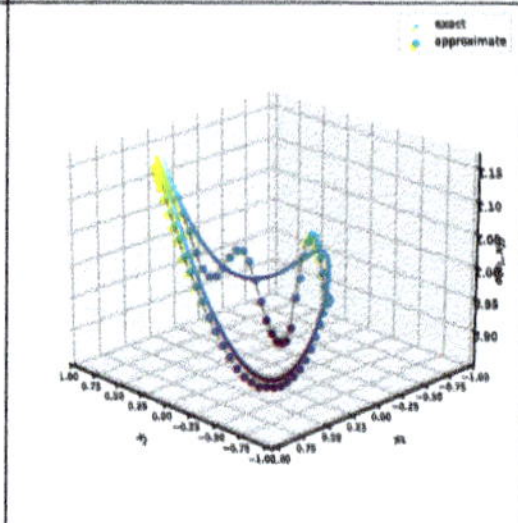	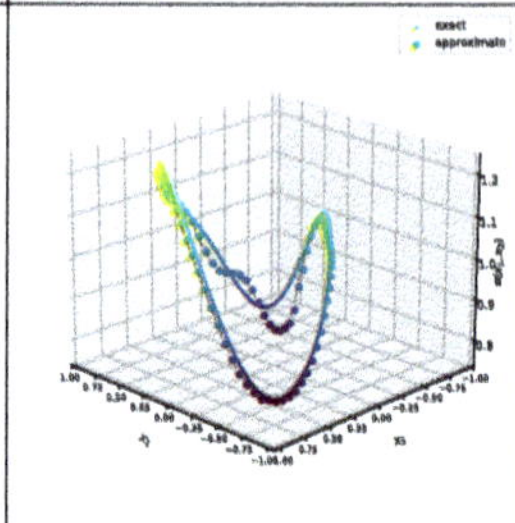
B2	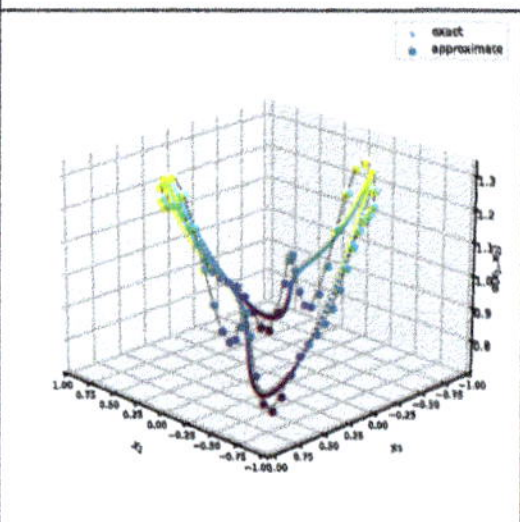	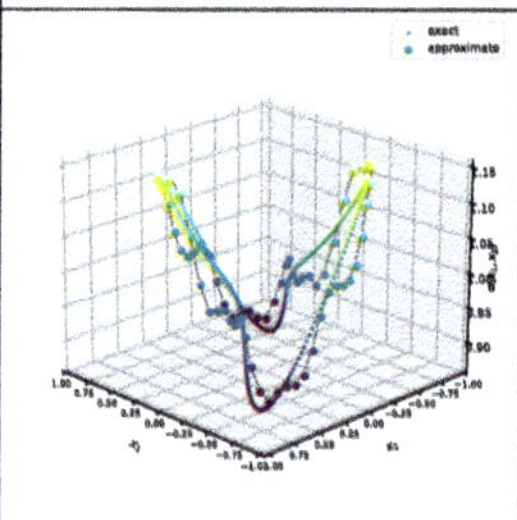	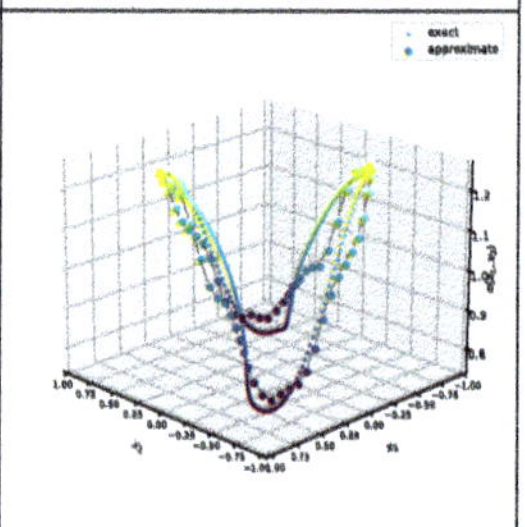
B3	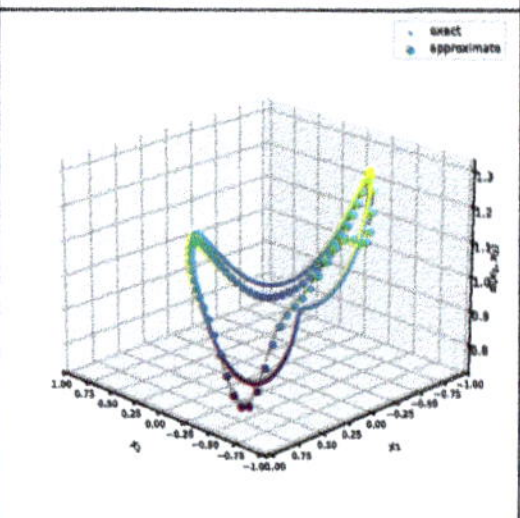	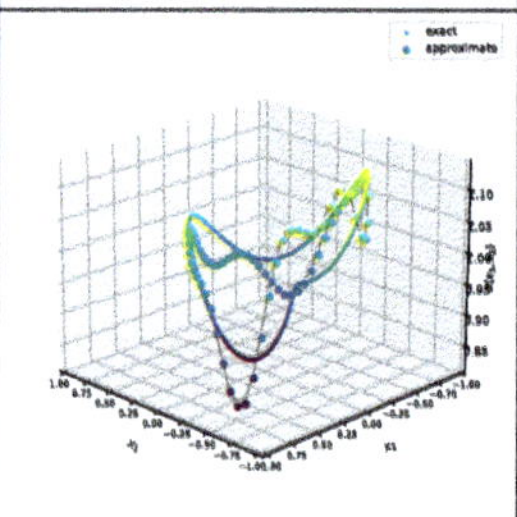	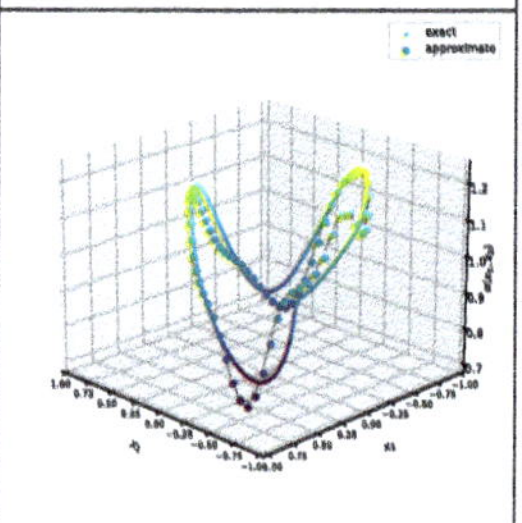

Tables 2 and 3. Even for complex shapes with concavities, the unknown boundary is reconstructed reliably despite 0.5% noise. However, the Robin coefficient reconstruction is poor due to the ill-posedness of the problem; see Tables 4 and 5. To address this issue, multiple measurements (more than two) are needed to obtain reasonable reconstruction of both the shape and the Robin coefficient; see, e.g., [30]. Figure 1 shows the histories of cost values, gradient norms, and balancing parameter τ. We see that the cost values are lower (as expected) under exact measurements and stabilize after several iterations for both exact and noisy data. Gradient norms also decrease significantly despite some oscillations.

Table 4 3D view of the reconstructed Robin coefficient, rotated by 30°, under 0.5% noise

Case	A1	A2	A3
B1	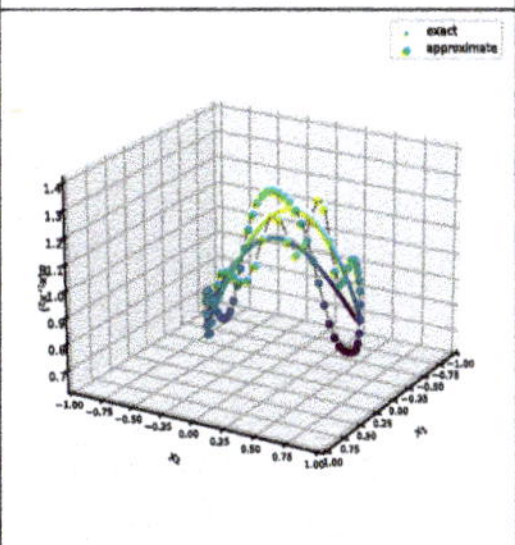	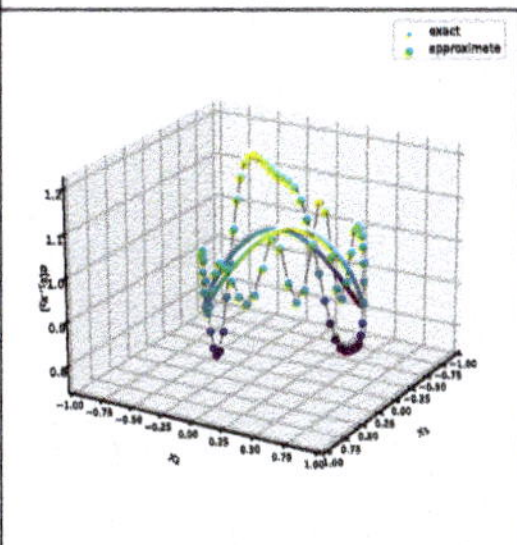	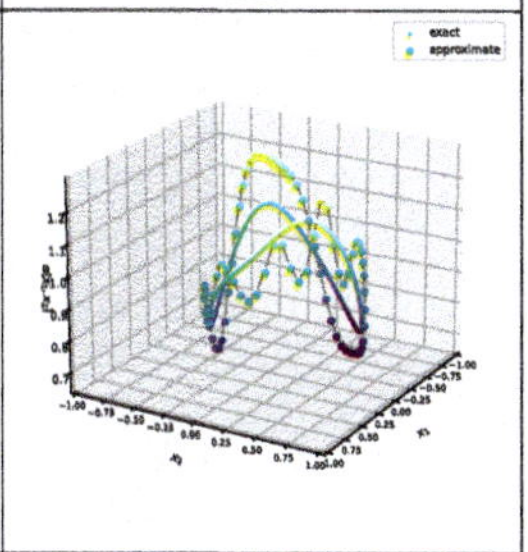
B2	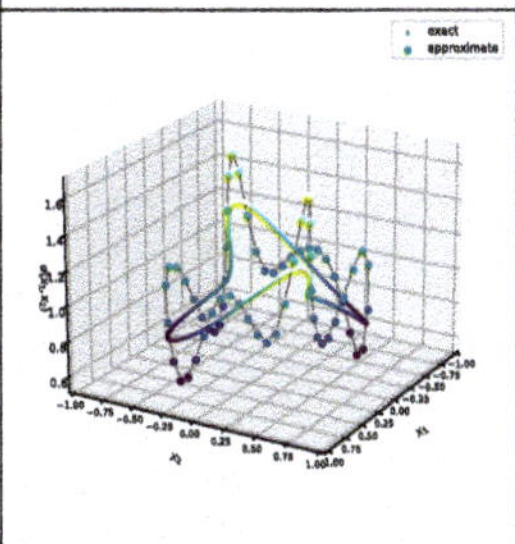	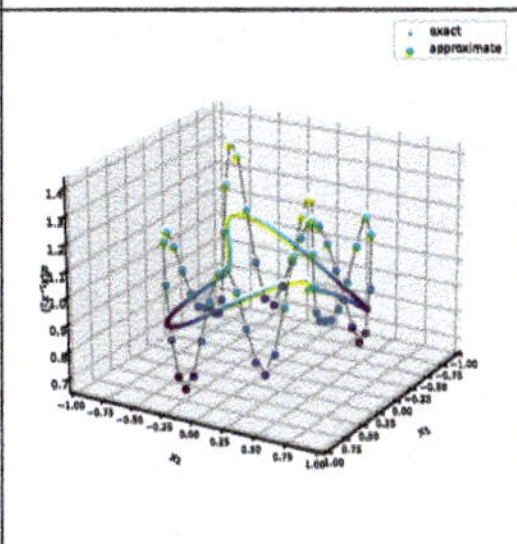	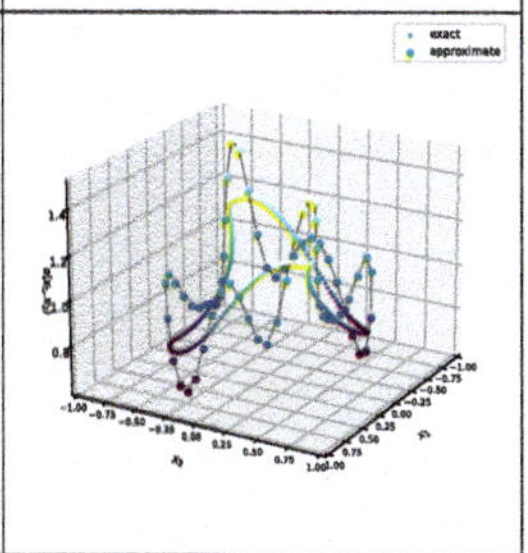
B3	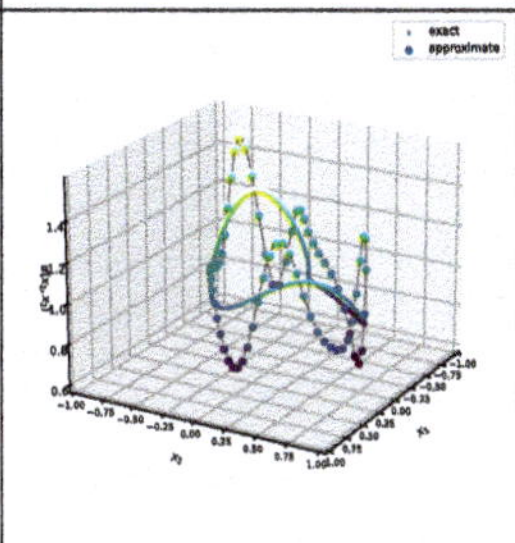	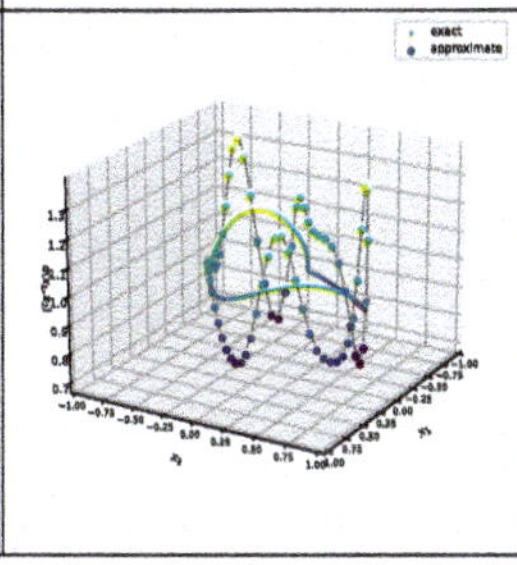	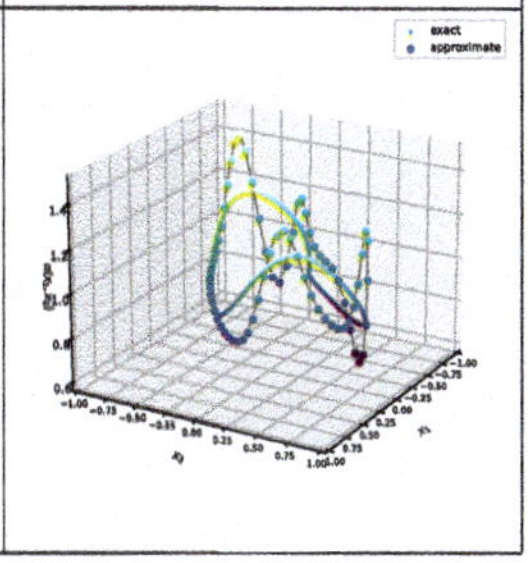

Overall, the method is effective under exact measurements, even for inclusions with concavities. Additional tests with varied initial geometries and inclusions (not shown) confirm consistent performance with exact data. For noisy data, using multiple measurements is essential for reasonable reconstructions. Improving noisy reconstructions will be the focus of future work.

Table 5 3D view of the reconstructed Robin coefficient, rotated by 135°, under 0.5% noise

Case	A1	A2	A3
B1	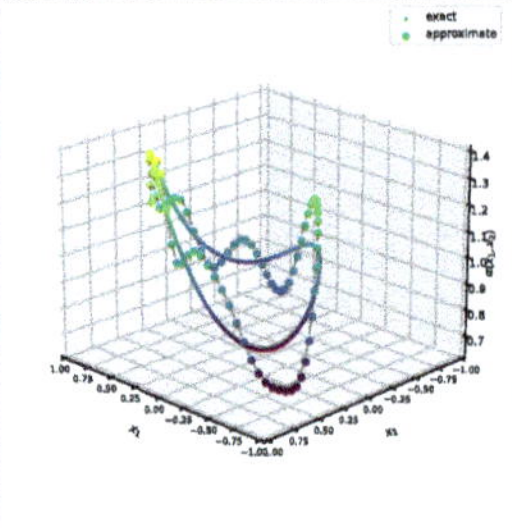	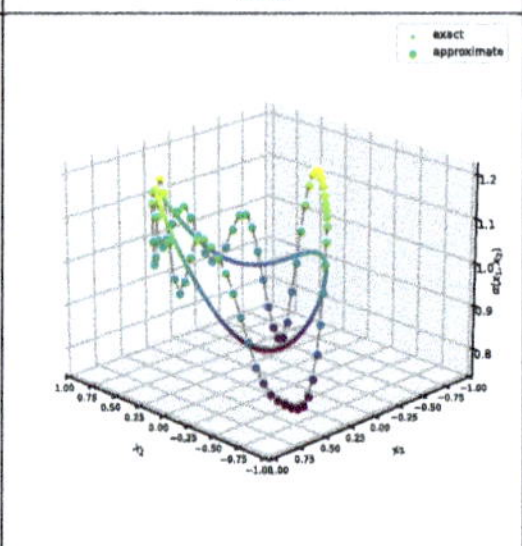	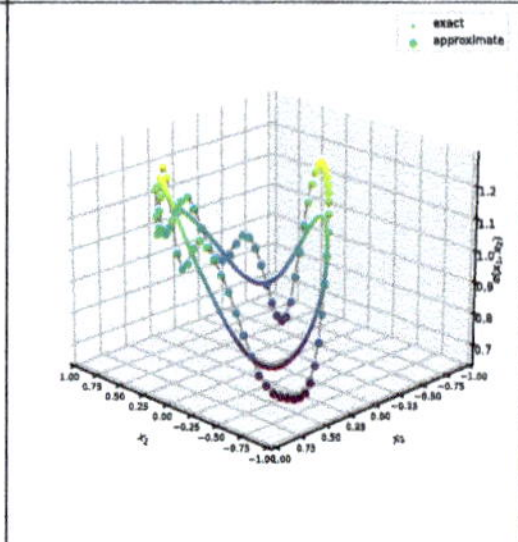
B2	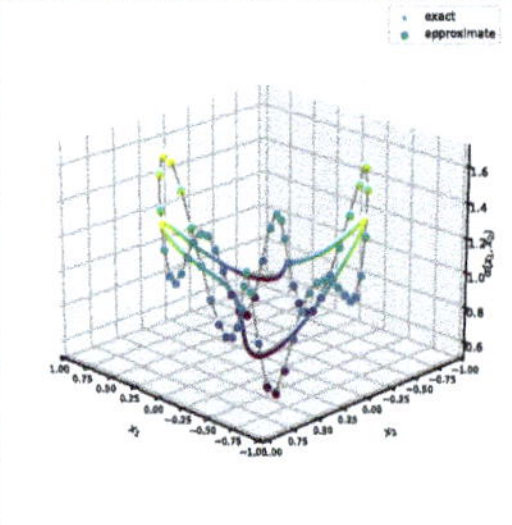	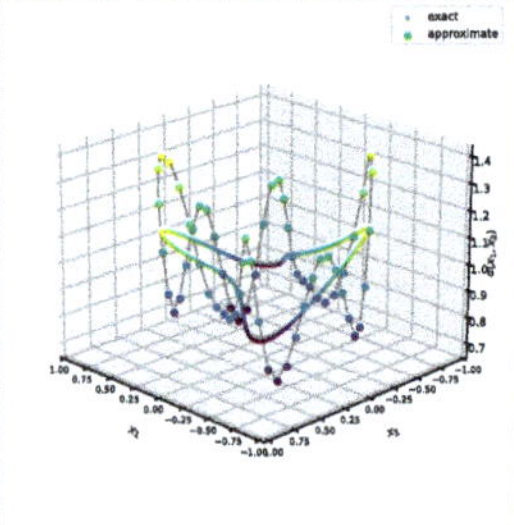	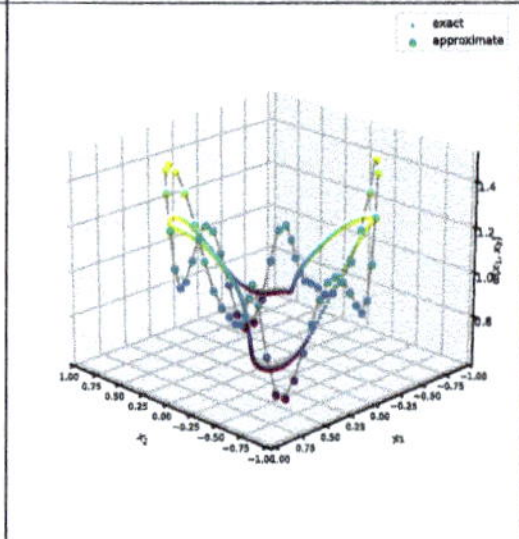
B3	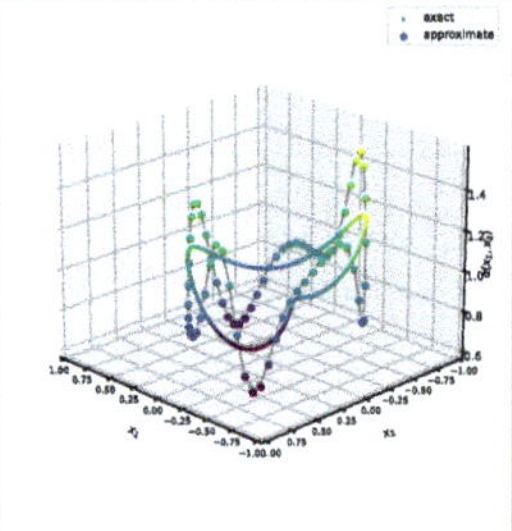	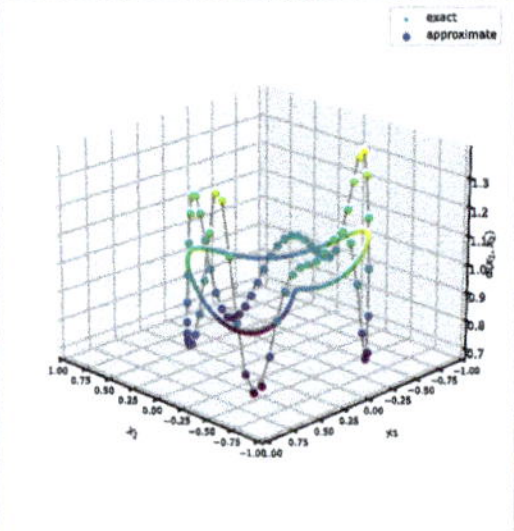	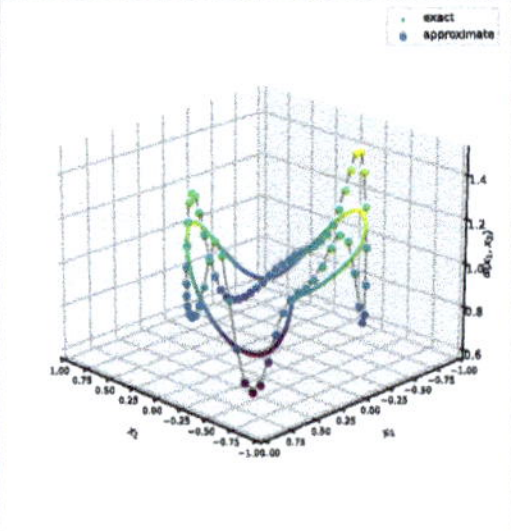

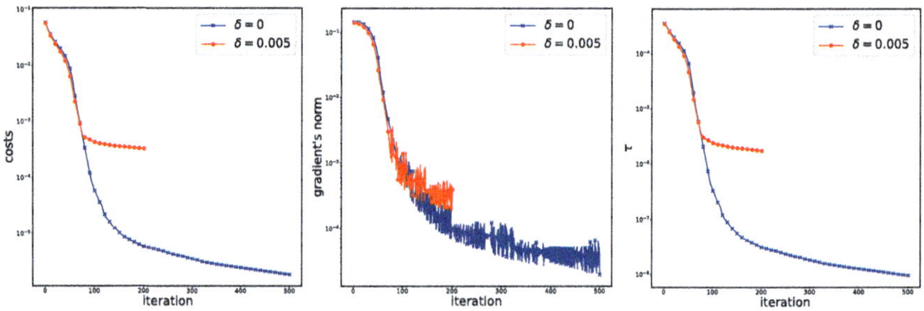

Fig. 1 Histories of cost values, gradient norms, and parameter τ for Case A3-B3

4 Conclusion

In this study, we presented a numerical method using the Kohn–Vogelius framework to simultaneously reconstruct an unknown inclusion and the Robin coefficient associated with an electrostatic problem. By utilizing gradient information derived from the energy-gap cost functional with respect to both the inclusion's shape and the Robin parameter, we formulated an iterative algorithm implemented via the finite element method. Numerical experiments demonstrated the method's feasibility and effectiveness for various inclusion geometries and Robin coefficients, provided exact measurements were available. For noisy data, however, incorporating multiple Cauchy pairs appears essential to achieve accurate reconstructions. Future work will address this aspect and be documented in a subsequent study.

Acknowledgements JFTR is supported by the JSPS Postdoctoral Fellowships for Research in Japan and partially by the JSPS Grant-in-Aid for Early-Career Scientists under Japan Grant Number JP23K13012 and the JST CREST Grant Number JPMJCR2014.

References

1. Ben Abda, A., Bouchon, F., Peichl, G.H., Sayeh, M., Touzani, R.: A Dirichlet-Neumann cost functional approach for the Bernoulli problem. J. Eng. Math. **81**, 157–176 (2013)
2. Afraites, L., Dambrine, M., Kateb, D.: Shape methods for the transmission problem with a single measurement. Numer. Funct. Anal. Optim. **28**(5–6), 519–551 (2007)
3. Afraites, L., Dambrine, M., Kateb, D.: On second order shape optimization methods for electrical impedance tomography. SIAM J. Control Optim. **47**(3), 1556–1590 (2008)
4. Afraites, L., Masnaoui, C., Nachaoui, M.: Shape optimization method for an inverse geometric source problem and stability at critical shape. Discrete Contin. Dyn. Syst. Ser. S **15**(1), 1–21 (2022)
5. Afraites, L., Rabago, J.F.T.: Boundary shape reconstruction with robin condition: existence result, stability analysis, and inversion via multiple measurements. Comput. Appl. Math. **43**(270), 37 (2024)
6. Afraites, L., Rabago, J.: Shape optimization methods for detecting an unknown boundary with the Robin condition by a single measurement. Discrete Contin. Dyn. Syst. Ser. S **18**(1), 43–76 (2025)
7. Azegami, H.: A solution to domain optimization problems. Trans. Jpn. Soc. Mech. Eng. Ser. A. **60**, 1479–1486 (1994)
8. Azegami, H.: Shape Optimization Problems. Springer Optimization and Its Applications, vol. 164. Springer, Singapore (2020)
9. Bacchelli, V.: Uniqueness for the determination of unknown boundary and impedance with the homogeneous Robin condition. Inverse Probl. 25:Art. 015004 (4pp) (2009)
10. Bacani, J.B., Peichl, G.H.: On the first-order shape derivative of the Kohn-Vogelius cost functional of the Bernoulli problem. Abstr. Appl. Anal. **2013**, 19 pp. Article ID 384320 (2013)
11. Caubet, F., Dambrine, M., Kateb, D., Timimoun, C.Z.: A Kohn-Vogelius formulation to detect an obstacle immersed in a fluid. Inverse Prob. Imaging **7**(1), 123–157 (2013)
12. Chaabane, S., Elhechmi, C., Jaoua, M.: A stable recovery method for the Robin inverse problem. Math. Comput. Simul. **66**, 367–383 (2004)
13. Chaabane, S., Jaoua, M.: Identification of Robin coefficients by means of boundary measurements. Inverse Problems **15**, 1425–1438 (1999)

14. Clason, C., Jin, B.: A semismooth Newton method for nonlinear parameter identification problems with impulsive noise. SIAM J. Imaging Sci. **5**, 505–536 (2012)
15. Clason, C., Jin, B., Kunisch, K.: A duality-based splitting method for l^1-tv image restoration with automatic regularization parameter choice. SIAM J. Sci. Comput. **32**, 1484–1505 (2010)
16. Clason, C., Jin, B., Kunisch, K.: A semismooth newton method for l^1 data fitting with automatic choice of regularization parameters and noise calibration. SIAM J. Imaging Sci. **3**, 199–231 (2010)
17. Cakoni, F., Kress, R.: Integral equations for inverse problems in corrosion detection from partial cauchy data. Inverse Prob. Imaging **1**, 229–245 (2007)
18. Cakoni, F., Kress, R., Schuft, C.: Integral equations for inverse problems in corrosion detection from partial Cauchy data. Inverse Probl. **26**, Art. 095012 24pp. (2010)
19. Cakoni, F., Kress, R., Schuft, C.: Simultaneous reconstruction of shape and impedance in corrosion detection. Methods Appl. Anal. **17**, 357–378 (2010)
20. Clason, C.: l^∞ fitting for inverse problems with uniform noise. Inverse Probl. **28**, 104007 (2012)
21. Chakib, A., Nachaoui, A., Nachaoui, M.: Existence analysis of an optimal shape design problem with non coercive state equation. Nonlinear Anal.: Real World Appl. **28**, 171–183 (2016)
22. Doğan, G., Morin, P., Nochetto, R.H., Verani, M.: Discrete gradient flows for shape optimization and applications. Comput. Methods Appl. Mech. Engrg. **196**, 3898–3914 (2007)
23. Eppler, K., Harbrecht, H.: A regularized Newton method in electrical impedance tomography using shape hessian information. Control Cybern. **34**, 203–225 (2005)
24. Eppler, K., Harbrecht, H.: On a Kohn-Vogelius like formulation of free boundary problems. Comput. Optim. App. **52**, 69–85 (2012)
25. Fang, W.: Simultaneous recovery of Robin boundary and coefficient for the Laplace equation by shape derivative. J. Comput. Appl. Math. **413**, Art. 114376 13 pp (2022)
26. Fasino, D., Inglese, G.: Recovering nonlinear terms in an inverse boundary value problem for Laplace's equation: a stability estimate. J. Comput. Appl. Math. **198**, 460–470 (2007)
27. Fang, W., Lu, M.: A fast collocation method for an inverse boundary value problem. Int. J. Numer. Methods Engrg. **59**, 1563–1585 (2004)
28. Fang, W., Lin, F., Ma, Y.: Fast algorithms for boundary integral equations on elliptic domains and related inverse problems. East Asian J. Appl. Math. **9**, 485–505 (2019)
29. Fang, W., Zeng, S.: Numerical recovery of Robin boundary from boundary measurements for the Laplace equation. J. Comput. Appl. Math. **224**, 573–580 (2009)
30. Harris, I.: A direct method for reconstructing inclusions and boundary conditions from electrostatic data. Appl. Anal. **102**(5), 1511–1529 (2023)
31. Hecht, F.: New development in FreeFem++. J. Numer. Math. **20**, 251–265 (2012)
32. Hauptmann, A., Ikehata, M., Itou, H., Siltanen, S.: Revealing cracks inside conductive bodies by electric surface measurements. Inverse Probl. **35**(025004), 24 (2019)
33. Hadri, A., Laghrib, A., Nachaoui, A., Nachaoui, M.: On the resolution of the non-smooth inverse cauchy problem by the primal-dual method. Numer. Algor. 1–34 (2024)
34. Hadri, A., Nachaoui, M., Laghrib, A., Chakib, A., Afraites, L.: A primal-dual approach for the robin inverse problem in a nonlinear elliptic equation: the case of the $l_1 - l_2$ cost functional. J. Inverse Ill-posed Probl. **31**(2), 243–266 (2023)
35. Henrot, A., Pierre, M.: Shape Variation and Optimization: A Geometrical Analysis. Tracts in Mathematics, vol. 28. European Mathematical Society, Zürich (2018)
36. Ito, K., Jin, B., Takeuchi, T.: A regularization parameter for nonsmooth tikhonov regularization. SIAM J. Sci. Comput. **33**, 1415–1438 (2011)
37. Inglese, G., Mariani, F.: Corrosion detection in conducting boundaries. Inverse Probl. **20**, 1207–1215 (2004)
38. Jin, B.: Conjugate gradient method for the robin inverse problem associated with the laplace equation. Internat. J. Numer. Methods Engrg. **71**, 433–453 (2007)
39. Kress, R., Colton, D.: Inverse Acoustic and Electromagnetic Scattering Theory. Applied Mathematical Sciences, vol. 93, 3rd edn. Springer, New York (1998)

40. Kaup, P.G., Santosa, F.: Nondestructive evaluation of corrosion damage using electrostatic measurements. J. Nondestr. Eval. **14**, 127–136 (1995)
41. Kaup, P.G., Santosa, F., Vogelius, M.: Method for imaging corrosion damage in thin plates from electrostatic data. Inverse Probl. **12**, 279–293 (1996)
42. Kohn, R., Vogelius, M.: Relaxation of a variational method for impedance computed tomography. Commun. Pure Appl. Math. **40**(6), 745–777 (1987)
43. Lin, F., Fang, W.: A linear integral equation approach to the robin inverse problem. Inverse Probl. **21**, 1757–1772 (2005)
44. Meftahi, H.: Uniqueness, Lipschitz stability, and reconstruction for the inverse optical tomography problem. SIAM J. Control Optim. **53**(6), 6326–6354 (2021)
45. Machida, M., Notsu, H., Rabago, J.F.T.: Reconstruction approach for optical tomography with fluorophores by single boundary measurement. In preparation (2024)
46. Neuberger, J.W.: Sobolev Gradients and Differential Equations. Springer, Berlin (1997)
47. Pagani, C.D., Peerotti, D.: Identifiability problems of defects with Robin condition. Inverse Probl. **25**, Art. 055007 12pp (2009)
48. Rabago, J., Azegami, H.: Shape optimization approach to defect-shape identification with convective boundary condition via partial boundary measurement. Japan J. Indust. Appl. Math. **31**(1), 131–176 (2018)
49. Rabago, J., Azegami, H.: A new energy-gap cost functional cost functional approach for the exterior Bernoulli free boundary problem. Evol. Equ. Control Theory **8**(4), 785–824 (2019)
50. Rabago, J., Azegami, H.: A second-order shape optimization algorithm for solving the exterior Bernoulli free boundary problem using a new boundary cost functional. Comput. Optim. Appl. **77**(1), 251–305 (2020)
51. Rundell, W.: Recovering an obstacle and its impedance from Cauchy data. Inverse Probl. **24**, 1–22 (2008)
52. Sincich, E.: Stability for the determination of unknown boundary and impedance with a Robin boundary condition. SIAM J. Math. Anal. **42**, 2922–2943 (2010)

A Primal-Dual Approach to Fractional Optimal Control in Non-smooth Machine Learning: FODE-Based Solutions for Enhanced Image Denoising

Fakhr-eddine Limami, Amine Laghrib, and Lekbir Afraites

Abstract This paper explores the development of advanced neural network architectures for image denoising. By integrating fractional calculus into deep convolutional neural networks, we introduce an optimal control framework, where network weights are iteratively optimized using Primal-Dual algorithm. This approach yields a highly efficient and faster numerical algorithm for image denoising. To validate the method's effectiveness, we showcase experimental results demonstrating superior performance compared to existing techniques.

Keywords Primal-Dual · Fractional optimal control · Neural networks · Fractional ordinary differential equation

1 Introduction

Digital imaging devices are widely used across various fields, such as remote sensing, facial recognition, and medical imaging. However, images captured by these devices often degrade in quality due to several factors, including noise interference and inconsistent lighting. In particular, the degradation of the underlying "true" or latent image occurs throughout processes like transmission and compression, where noise is introduced. To remove this noise and accurately recover the latent image from a degraded one, image denoising techniques are essential. Image denoising has attracted significant attention in recent years as it plays a crucial role in enhancing image quality in the field of image processing. The objective is to reconstruct a clean image from a noisy input, where a noisy image can typically be modeled as

$$\mathcal{U} = u + \eta, \tag{1}$$

F. Limami (✉) · L. Afraites
EMI FST Béni-Mellal, Université Sultan Moulay Slimane, Beni Mellal, Morocco
e-mail: fakhr.limami@gmail.com

A. Laghrib
EMI ENSA Khouribga, Université Sultan Moulay Slimane, Beni Mellal, Morocco

A. Laghrib et al. (eds.), *Computational Methods for Inverse Problems and Applications*, Springer Proceedings in Mathematics & Statistics 498,
https://doi.org/10.1007/978-3-031-89498-5_7

with $\mathcal{U}$ being the observed noisy image, u the desired clean image, and η the noise component. Denoising methods may be roughly divided into two categories: learning-based methods and model-based methods. Both aim to lower image noise, but they accomplish it using different frameworks and techniques. To address the denoising problem, model-based techniques rely on exacting mathematical models, such as partial differential equations (PDEs) [1, 2]. These techniques are based on a knowledge of image structures and noise properties, which enables them to develop models that successfully eliminate noise while maintaining important image characteristics. Model-based denoising usually entails optimizing certain mathematical functions to get the optimum outcome, frequently utilizing optimization algorithms that take into account the characteristics of the image and the noise. Learning-based methods, particularly those that use supervised deep learning [3], use a different strategy by training deep convolutional neural networks (DCNNs) [4] on large datasets. Learning-based approaches allow the model to learn the denoising job directly from data, as opposed to explicitly creating a mathematical model. A DCNN can capture intricate, non-linear correlations between the noisy input and clean output by learning to map a noisy image to its matching clean version through training. This method may provide very good denoising models that can handle different kinds and structures of noise. Inspired by the work of [5, 6] who developed new neural network architectures successively named FOCNet and $\theta-$FOCNet, we deal with a problem of minimizing a cost function L^1 by adding a Tikhonov regularization under the constraint of a fractional ordinary differential equation. Next, we reformulate the optimization problem into a mini-max structure. To solve this, we choose the Primal-Dual method, known for its effectiveness in handling non-smooth optimization challenges [7]. This method can bypass the non-differentiability of the loss function by employing Fenchel conjugates and proximal operators. The structure of the paper is summarized below. In Sect. 2, we describe the issue setting and use the L^1 loss function to reformulate it as a minimization model. The Primal-Dual method and its usage in our optimization model are the main topics of Sect. 3. In Sect. 4, we evaluate the algorithm's stability and performance by contrasting it with certain denoising techniques found in the literature. We wrap up with a summary and discuss potential avenues for further study. The numerical results show that the suggested method is a useful tool for image denoising.

2 Proposed Model

A supervised approach is the main method used to train our model. The network is trained using paired instances of noisy images, $u(0, x)$, and their matching clean (denoised) counterparts, $v(x)$. The network seeks to minimize the difference between the denoised output and the real clean image in order to align the noisy inputs with their clean equivalents. During this learning phase, the network's parameters are changed in order to optimize this mapping. The supervised structure guarantees that the model learns to eliminate noise efficiently while maintaining important aspects of the image. The aim of the problem is to identify the optimal control that minimizes the objective loss.

$$\min_{\gamma \in \mathbb{S}_{ad}} ||u(T,x) - v(x)||_{L^1(\Omega)} + \frac{\lambda}{2}||\gamma(t)||^2_{L^2(0,T)}. \tag{2}$$

The equation of state for the problem, which starts with the noisy input $u(0,x)$ and follows its evolution over time, specifying the complete denoising process is given by:

$$\begin{cases} {}^c\mathfrak{D}_t^\alpha u(t,x) = f(u(t,x),\gamma(t)),\ (t,x) \in (0,T) \times \Omega \\ u(0,x) = a(x). \end{cases} \tag{3}$$

Here, $\Omega \subset \mathbb{R}^2$ has a Lipschitz continuous boundary $\partial\Omega$; a represents the observed noisy image input, and v denotes the ground truth clean image. The constant λ is a regularization parameter. The term ${}^c\mathfrak{D}_t^\alpha u$ represents the left Caputo derivative of u of order α, defined as

$${}^c\mathfrak{D}_t^\alpha u(t) := \frac{1}{\Gamma(1-\alpha)} \int_0^t \frac{u'(\tau)}{(t-\tau)^\alpha} d\tau, \tag{4}$$

with $\alpha \in (0,1)$ and Γ is Euler's Gamma function. The set $\mathbb{S}_{ad}$ defines the set of admissible controls as

$$\mathbb{S}_{ad} := \left\{\gamma \in W^{1,\infty}(0,T) \mid \gamma \text{ is Lipschitzian}\right\}.$$

The choice of the equation of state as a differential equation is linked to the fact that forward propagation from one layer to another in a neural network can be modelled in the form of a differential equation thanks to the dynamic nature of the propagation. In the literature, several architectures are linked to this choice, such as RESNet [8]. On the other hand, we have chosen a fractional differential equation of state of order $0 < \alpha < 1$ to take advantage of the fact that it provides a non-local approach which results in a longer memory compared to equations of integer order.

Remark 1 The existence and uniqueness of the solution to the Eq. (3) is already covered in the article [6].

3 The Proposed Algorithm

In this research, we address the fractional optimal control problem by framing it as an inverse problem, employing a broad, non-smooth cost functional. Specifically, we utilize an L^1 loss function, which offers higher accuracy and favorable convergence properties, to represent and solve the problem. Numerous methods are examined in order to determine the minimizer of the suggested optimization issue (2). In order to handle non-smooth optimization problems, a modified version of the Primal-Dual method, which was first presented by Chambolle and Pock [9], has recently been created. The resultant optimization issue is first converted to a mini-max problem. We choose the Primal-Dual approach to address this problem since it has been demonstrated to be effective in handling non-smooth optimization problems [7]. Two convex problems are formulated as the basic idea underlying the Primal-Dual method, which is a technique used to address the problem's intrinsic non-regularity, because it may use the proximal operator and the Fenchel conjugate to get around the loss function's non-differentiability.
We consider the operator

$$\begin{aligned} \mathcal{T} :& \mathrm{S}_{ad} \to L^1(\Omega) \\ & \gamma \to u(T,x) - v(x), \end{aligned}$$

Consequently, the problem (2) may be expressed as follows:

$$\min_{\gamma \in \mathrm{S}_{ad}} \mathcal{F}(\mathcal{T}(\gamma)) + \mathcal{R}(\gamma), \tag{5}$$

with

$$\mathcal{F}(\mathcal{T}(\gamma)) = ||\mathcal{T}(\gamma)||_{L^1(\Omega)}, \tag{6}$$

and

$$\mathcal{R}(\gamma) = \frac{\lambda}{2}||\gamma(t)||^2_{L^2(0,T)}, \tag{7}$$

The generic saddle-point problem that follows is a formulation of problem (5)

$$\min_{\gamma \in \mathrm{S}_{ad}} \max_{\eta \in \mathrm{S}_{ad}} \mathcal{R}(\gamma) + \langle \mathcal{T}(\gamma), \eta \rangle - \mathcal{F}^*(\eta), \tag{8}$$

with $\mathcal{F}^*$ is the Fenchel conjugate operator of $\mathcal{F}$ defined by

$$\mathcal{F}^*(\eta) := \sup_c \langle c, \eta \rangle - \mathcal{F}(c) = i_A(d), \tag{9}$$

where $i_A(d)$ is the indicator function of the set A defined by:

$$i_A(d) = \begin{cases} 0, & if\ d \in A, \\ \infty, & otherwise. \end{cases}$$

And the set A defined by

$$A := \{\omega \in L^\infty \text{ such that } ||\omega||_\infty \leq 1\}.$$

The Primal-Dual algorithm's iterations may now be applied, yielding the following iterative algorithm.

$$\begin{cases} \gamma^{n+1} = (I + \mu\partial\mathcal{R})^{-1}(\gamma^n - \mu(\mathcal{T}'(\gamma^n))^*\eta^n), \\ \hat{\gamma}^{n+1} = \gamma^{n+1} + \kappa(\gamma^{n+1} - \gamma^n), \\ \eta^{n+1} = (I + \sigma\partial\mathcal{F}^*)^{-1}(\eta^n + \sigma\mathcal{T}(\hat{\gamma}^{n+1})). \end{cases}$$

With $\sigma, \kappa, \mu \in \mathbb{R}^{+*}$, now we will compute the proximal operators $(I + \mu\partial\mathcal{R})^{-1}$ and $(I + \sigma\partial\mathcal{F}^*)^{-1}$.
The proximal operator of $\mathcal{R}$ is defined by:

$$(I + \mu\partial\mathcal{R})^{-1}(\hat{\gamma}) = \mathbf{prox}_{\mu\mathcal{R}}(\hat{\gamma}) = \arg\min_{\omega} \frac{||\omega - \hat{\gamma}||_2^2}{2\mu} + \mathcal{R}(\gamma), \tag{10}$$

by the fact that $\mathcal{R}(\gamma) = \frac{\lambda}{2}||\gamma||_2^2$, we get

$$\mathbf{prox}_{\mu\mathcal{R}}(\hat{\gamma}) = (I + \lambda\mu)^{-1}\hat{\gamma}, \tag{11}$$

in other hand, the projection on a convex set A may be used to define the operator $(I + \sigma\partial\mathcal{F}^*)^{-1}$ as follows

$$(I + \sigma\partial\mathcal{F}^*)^{-1}(\hat{\eta}) = \mathbf{prox}_{\sigma\mathcal{F}}(\hat{\eta}) = \mathbf{Proj}_A(\hat{\eta}) = \frac{\hat{\eta}}{\max(||\hat{\eta}||_\infty, 1)}, \tag{12}$$

to complete the iterative Primal-Dual algorithm, the adjoint derivative of $\mathcal{T}$ is calculated using the adjoint state associated with (3), which gives the following result

$$\mathcal{T}'(\gamma) = \int_\Omega f'(u, \gamma) p dx, \tag{13}$$

with p is the solution of the following problem adjoint:

$$\begin{cases} {}^{c}\mathfrak{D}^{\alpha}_{T-t} p(t,x) + u'(t,x)\delta_T(t) - f'(u,\gamma)p(t,x) = 0 \\ p(T) = 0. \end{cases} \tag{14}$$

where $\delta_T(t)$ is a Dirac function. Now, we apply the Primal-Dual algorithm associated with problem (2) as follow:

Algorithm 1 Accelerated Primal-Dual Algorithm

Input: $T > 0$, ζ, $\alpha \in (0,1)$, λ, the function f and the initial condition $u^0 = a$. Choose $\mu^0 > 0$, $\sigma^0 > 0$, κ^0 and ϵ.
Initialization: η^0, set $z^k = \gamma_0^k$ and $n = 0$.
Begin:
for: $k = 0, 1, 2, \dots, T$ **do:**

- Compute u_i^k from (3).
- Compute p_i^k from (14).
- Compute $(\mathcal{T}')^*(\gamma_n^k).\eta^n$ from (13).
- $\gamma_{n+1}^k = (I + \mu^n \partial \mathcal{R})^{-1}(\gamma_n^k - \mu^n(\mathcal{T}'(\gamma_n^k))^*\eta^n)$ from (11).
- $\kappa^n = \frac{1}{\sqrt{2\zeta\sigma^n+1}}$; $\mu^{n+1} = \frac{\mu^n}{\kappa^n}$; $\sigma^{n+1} = \sigma^n\kappa^n$.
- $z_{n+1}^k = \gamma_{n+1}^k + \kappa^n(\gamma_{n+1}^k - \gamma_n^k)$.
- $\eta^{n+1} = (I + \sigma^{n+1}\partial\mathcal{F}^*)^{-1}(\eta^n + \sigma^{n+1}\mathcal{T}(z_{n+1}^k))$ from (12).
- n=n+1.

End for
Iterate until: $\frac{||\mathcal{T}(\gamma_{n+1}^k) - \mathcal{T}(\gamma_n^k)||_2}{||\mathcal{T}(\gamma_{n+1}^k)||_2}$
Output: $\gamma^{k+1} = \gamma_{n+1}^k$.

4 Numerical Experiments

In this section, we present selective results of image denoising to demonstrate the effectiveness of our proposed non-convex algorithm for impulse noise removal. To highlight the advantages of our approach, we compare our denoising results with those of several state-of-the-art impulse noise removal methods.

For dataset generation, we use the Set12 dataset, from which we create training pairs of clean and noisy images (x_i, f_i). Specifically, we select some grayscale image patches of size 128×128 as ground truth clean images, and construct the noisy observations f_i by adding Gaussian white noise at various levels of $\sigma = 10, 30, 50$ to each clean patch.

To train our network, we employ the proposed Primal Dual algorithm, as opposed to the ADAM optimization method used by other state-of-the-art methods. The Primal Dual algorithm is used to minimize the Mean Squared Error (MSE) loss function, which includes a regularization term to optimize the parameters of our γ-FOCNet.

Regarding the Primal Dual parameters, we set the following: the feature maps for each layer are fixed at 50, and the network consists of 20 blocks. All experiments are conducted on an Nvidia Titan X GPU using the open-source TensorFlow framework. Additionally, we train one specific denoising network for each noise level.

For performance evaluation, we use the Berkeley Segmentation Dataset for testing. We consider two subsets: one containing grayscale images of size 512×512 and another with 256×256 color images. The architecture of our fractional network includes two hidden layers, each with 700 hidden units, and we use the Fast Rectified Linear Unit (FRELU) activation function in both network configurations.

We begin with a simulated test to evaluate the performance of our approach under controlled conditions. In this test, we apply our denoising method to a set of synthetic images that have been degraded with different levels of noise. The goal is to assess how well the proposed algorithm can restore the original image quality by comparing the restored images with their clean counterparts. For this purpose, we introduce a mixture of randomize Poisson noise and Gaussian noise of varying variance ($\sigma = 40, 45, 50, 55$) to the images, which allows us to simulate realistic noise levels typically encountered in image processing tasks. These results are presented in Fig. 1, where we observe that our method outperforms well techniques in terms of visual fidelity, particularly in preserving fine details and edges in the restored images.

To further assess the effectiveness of our approach, we compare the denoising results with those from several state-of-the-art methods, including: Tensor-guided learning (TGL) [1], the Nonlocal Laplace (NLL) [10], the non-convex PDE-constrained diffusion equation (NNDE) [2], and the variational PDNet network (VPDNet) [3].

For this evaluation, we apply the denoising algorithms to two selected tests *Fly* and *Cameraman* test images, which has a resolution of 512×512 and 256×256 pixels, respectively. The image is intentionally degraded by adding a randomize Poisson noise and a Gaussian noise with variances of $\sigma^2 = 0.04$ and $\sigma^2 = 0.05$. In Figure 2, we present the noisy input image, the result of denoising using our approach for the *Fly* image, and the outputs of the other methods for comparison. Also, in Fig. 3, we present the recovered *Cameraman* using the compared methods and also our proposed one. Our proposed method delivers subtle yet significant improvements in the restored image, particularly excelling in edge preservation and enhancing overall sharpness, which are crucial for maintaining image detail during the denoising process.

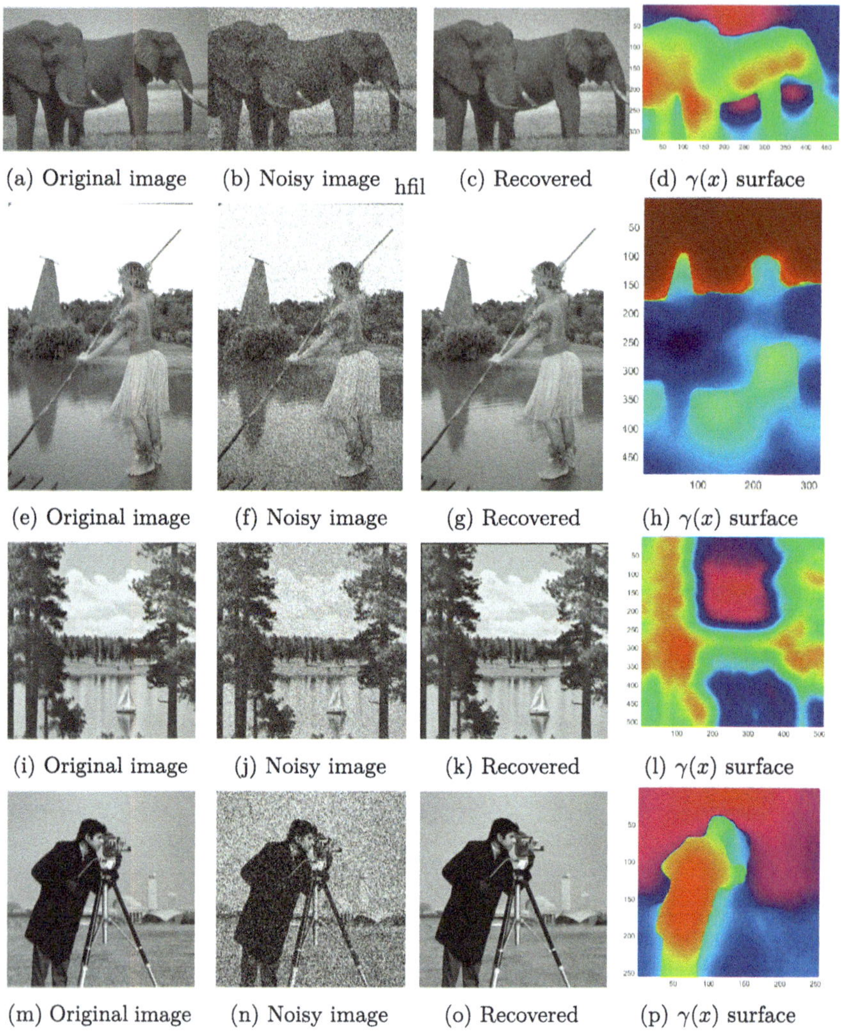

Fig. 1 The denoising result for the four selected images using various choice of σ noise and also the computed γ for each test

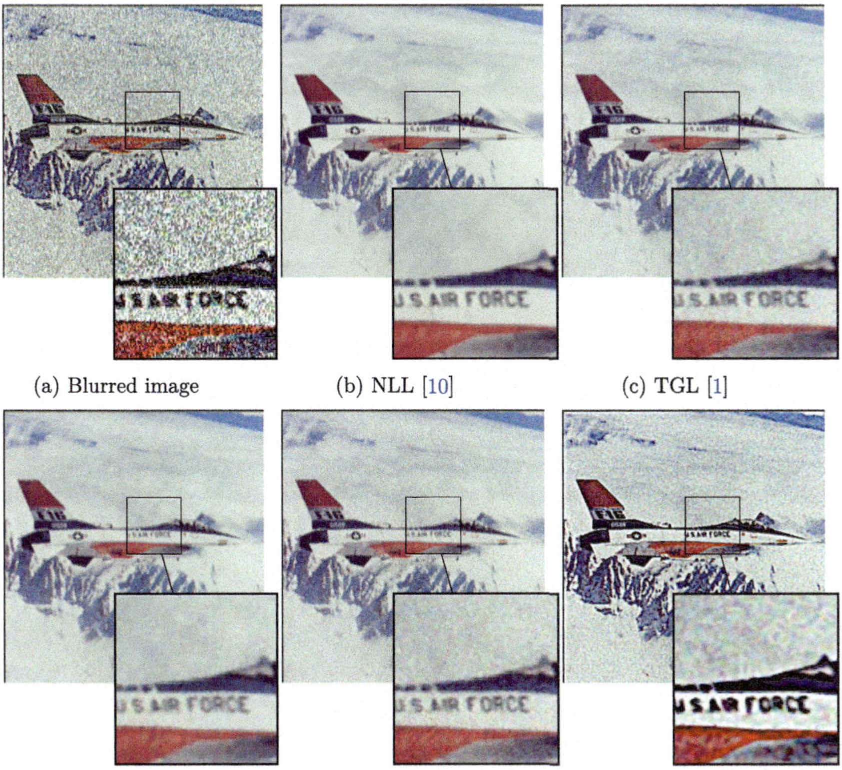

(a) Blurred image (b) NLL [10] (c) TGL [1]

(d) NNDE [6] (e) VPDNet [3] (f) Our model

Fig. 2 The restored image of (*Airplane*) using different denoising approaches

We conduct a quantitative assessment using the PSNR and SSIM metrics to provide objective evidence of the effectiveness and efficiency of our model with the variable γ. This analysis demonstrates the model's superiority in image reconstruction and restoration compared to other state-of-the-art methods. The PSNR and SSIM values for all methods are presented in Table 1. From our analysis, it is clear that the integration of the constrained problem consistently delivers better performance, outperforming existing methods in terms of both PSNR and SSIM scores. We conclude

(a) Blurred image (b) NLL [10] (c) TGL [1]

(d) NNDE [6] (e) VPDNet [3] (f) Our model

Fig. 3 The restored image of (*Cameraman*) using different denoising approaches

the experimental results with a final real-world test using the *Disk* image, which are depicted in Fig. 4. This test further demonstrates the practical effectiveness of our proposed approach in handling real image data, providing additional insights into its performance beyond synthetic scenarios. The image was subject to real noise conditions, and the results are compared to those obtained from other state-of-the-art denoising methods. The effectiveness of our model is evident in the enhanced quality of the restored image, showcasing superior denoising capabilities and preservation of fine details.

Table 1 The table of PSNR and SSIM values for additional denoising tests. The highest value in each row is bolded

Image	PSNR (SSIM)				
	NLL [10]	TGL [1]	NNDE [2]	VPDNet [3]	Proposed approach
Flower (4×4 and $\sigma = 15$)	32.11 (0.921)	32.39 (0.926)	32.74 (0.933)	33.01 (0.937)	33.17 (0.942)
Flower (8×8 and $\sigma = 25$)	28.54 (0.835)	28.97 (0.845)	29.35 (0.856)	29.76 (0.869)	29.84 (0.872)
Flower (10×10 and $\sigma = 35$)	25.91 (0.721)	26.14 (0.725)	26.52 (0.734)	26.91 (0.747)	27.05 (0.752)
Mountain (6×6 and $\sigma = 10$)	31.88 (0.900)	32.10 (0.902)	32.45 (0.911)	32.72 (0.913)	32.94 (0.915)
Mountain (10×9 and $\sigma = 20$)	29.67 (0.878)	29.95 (0.883)	30.21 (0.889)	30.55 (0.895)	30.72 (0.898)
Mountain (12×10 and $\sigma = 30$)	27.14 (0.811)	27.46 (0.815)	27.72 (0.820)	27.95 (0.824)	28.03 (0.826)
Car (5×5 and $\sigma = 10$)	34.72 (0.940)	34.88 (0.945)	35.02 (0.949)	35.29 (0.952)	35.46 (0.955)
Car (8×8 and $\sigma = 20$)	30.58 (0.880)	30.91 (0.885)	31.14 (0.891)	31.46 (0.894)	31.68 (0.897)
Car (10×10 and $\sigma = 30$)	27.03 (0.802)	27.37 (0.807)	27.56 (0.812)	27.79 (0.815)	27.94 (0.818)
Ship (6×6 and $\sigma = 15$)	32.29 (0.915)	32.48 (0.918)	32.85 (0.922)	33.05 (0.924)	33.23 (0.928)
Ship (9×9 and $\sigma = 25$)	28.94 (0.837)	29.17 (0.843)	29.44 (0.850)	29.73 (0.853)	29.85 (0.856)
Ship (12×12 and $\sigma = 40$)	25.58 (0.710)	25.92 (0.718)	26.20 (0.724)	26.47 (0.729)	26.58 (0.731)
Zebra (4×4 and $\sigma = 5$)	34.41 (0.957)	34.53 (0.960)	34.62 (0.963)	34.81 (0.965)	34.95 (0.968)
Zebra (6×6 and $\sigma = 15$)	29.81 (0.876)	30.03 (0.880)	30.22 (0.884)	30.45 (0.887)	30.58 (0.890)
Zebra (8×8 and $\sigma = 25$)	26.12 (0.742)	26.45 (0.748)	26.72 (0.753)	26.99 (0.757)	27.08 (0.760)

5 Conclusion

This paper presents a novel approach to image denoising by leveraging advanced neural network architectures and incorporating fractional calculus into deep convolutional networks. By framing the denoising task within an optimal control paradigm, we propose an innovative method where the network's weights are iteratively optimized using a Primal-Dual algorithm. The integration of fractional derivatives into the network structure enhances its ability to capture complex image features, providing a significant boost in denoising performance. Through extensive experimental validation, we demonstrate that the proposed method outperforms existing state-of-the-art denoising techniques, offering faster convergence times and superior restoration quality. These results underscore the potential of combining fractional calculus with deep learning to achieve more efficient and effective image denoising solutions.

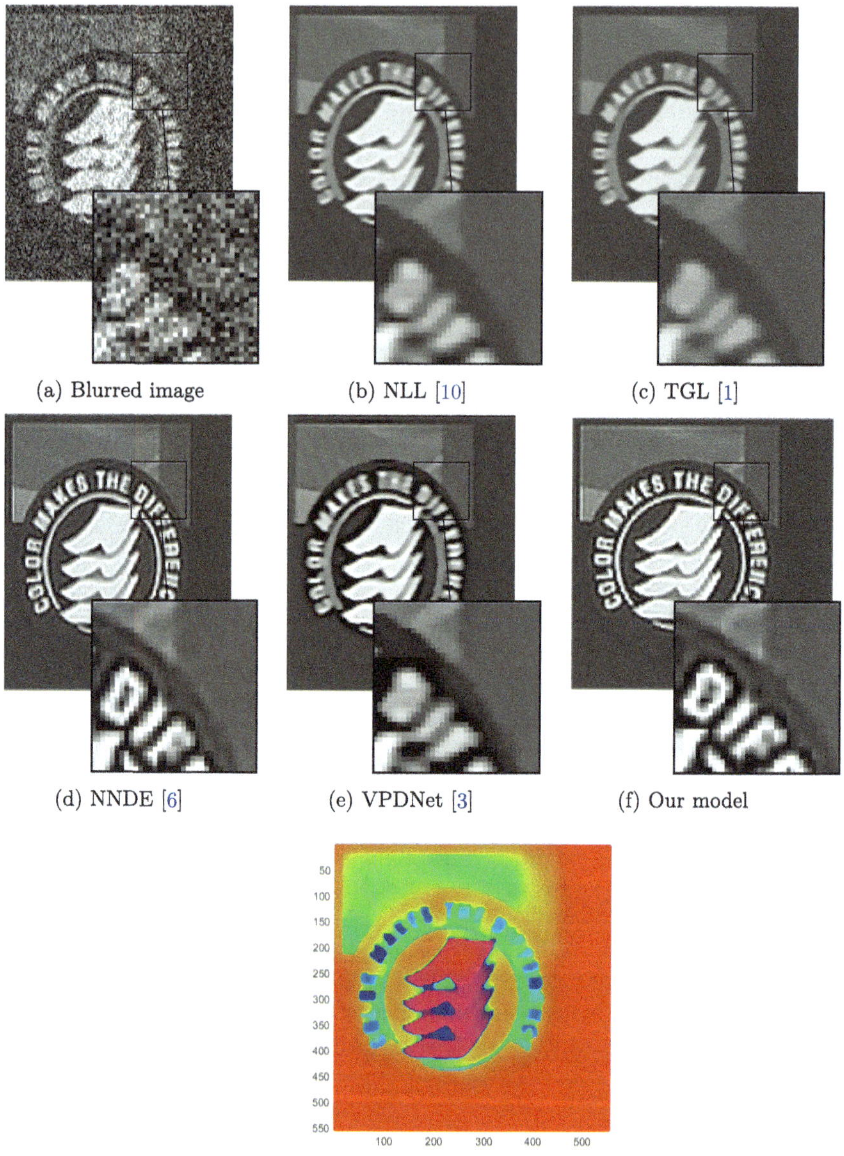

(a) Blurred image (b) NLL [10] (c) TGL [1]

(d) NNDE [6] (e) VPDNet [3] (f) Our model

(g) $\gamma(x)$ surface

Fig. 4 The restored image of the real (*Disk*) image using different denoising approaches

References

1. Limami, F.-E., Hadri, A., Afraites, L., Laghrib, A.: Tensor-guided learning for image denoising using anisotropic PDEs. Mach. Vis. Appl. **35**(3), 48 (2024). https://doi.org/10.1007/s00138-024-01532-4
2. Laghrib, A., Afraites, L., Hadri, A., Nachaoui, M.: A non-convex PDE-constrained denoising model for impulse and Gaussian noise mixture reduction. Inverse Probl. Imag. **17**(1), 23–67 (2023)
3. El Hakoume, A., Laghrib, A., Afraites, L., Hadri, A.: A variational PDNet network using a learning reaction-diffusion equation. Expert Syst. Appl. **249**, 123605 (2024)
4. Shin, H.-C., Roth, H.-R., Gao, M., Lu, L., Xu, Z., Nogues, I., Yao, J., Mollura, D., Summers, R.M.: Deep convolutional neural networks for computer-aided detection: CNN architectures, dataset characteristics and transfer learning. IEEE Trans. Med. Imag. **35**(5), 1285–1298 (2016)
5. Jia, X., Liu, S., Feng, X., Zhang, L.: FOCNet: A Fractional Optimal Control Network for Image Denoising, pp. 6054–6063 (2019). https://openaccess.thecvf.com/content_CVPR_2019/html/Jia_FOCNet_A_Fractional_Optimal_Control_Network_for_Image_Denoising_CVPR_2019_paper.html
6. Limami, F.-E., Hadri, A., Laghrib, A., Afraites, L.: Fractional optimal control for deep convolutional neural networks exploring ODE-based solutions for image denoising. Inverse Prob. Imag. (2024)
7. Lyaqini, S., Nachaoui, M., Hadri, A.: An efficient primal-dual method for solving non-smooth machine learning problem. Chaos, Solitons Fract. **155**, 111754 (2022)
8. Krizhevsky, A., Sutskever, I., Hinton, G.-E.: Imagenet classification with deep convolutional neural networks. Adv Neural Inf. Process. Syst. 25 (2012)
9. Chambolle, A., Pock, T.: A first-order primal-dual algorithm for convex problems with applications to imaging. J. Math. Imag. Vis. **40**(1), 120–145 (2011)
10. Laghrib, A., Ghazdali, A., Hakim, A., Raghay, S.: A multi-frame super-resolution using diffusion registration and a nonlocal variational image restoration. Comput. Math. Appl. **72**(9), 2535–2548 (2016)

Regularization and Solution of the Cauchy Problem in Anisotropic Heat Conduction

Hafida Hamdi, Mohamed Ziouane, Abdeljalil Nachaoui, and Mourad Nachaoui

Abstract This work addresses inverse problems encountered in the mathematical modeling of thermal conduction, with a focus on the Cauchy problem, a classic example of an ill-posed inverse problem. Due to uncertainties in conditions and parameters, reconstructing inaccessible information from partial data is often unstable and computationally expensive. As a result, a regularization approach is essential. We propose a numerical method to stabilize the problem by reformulating it as an optimal control problem with regularization. To overcome the limitations of first-order methods and efficiently handle large-scale data, we introduce the Limited Memory BFGS optimization algorithm to ensure stability and convergence. Numerical experiments in two dimensions are performed using the finite element method to approximate the direct and adjoint problems at each iteration.

Keywords Inverse problem · Cauchy problem · Control problem · Finite element · Limited memory BFGS

1 Introduction

Partial differential equation (PDE) modeling is crucial in real-life applications [1–3]. Indeed, due to their ability to describe and predict how quantities change over space and time, PDEs are commonly used to understand and address complex phenomena

---

Mohamed Ziouane, Abdeljalil Nachaoui, Mourad Nachaoui these authors contributed equally to this work.

H. Hamdi (✉)
EMI, Faculté des Sciences et Techniques, Université Sultan Moulay Slimane, Béni-Mellal, Morocco
e-mail: hafida.hamdi@usms.ma

M. Ziouane · A. Nachaoui · M. Nachaoui
Laboratoire de Mathématiques Jean Leray UMR6629 CNRS, Université de Nantes, Nantes, France
e-mail: abdeljalil.nachaoui@univ-nantes.fr

M. Nachaoui
e-mail: m.nachaoui@usms.ma

A. Laghrib et al. (eds.), *Computational Methods for Inverse Problems and Applications*, Springer Proceedings in Mathematics & Statistics 498,
https://doi.org/10.1007/978-3-031-89498-5_8

across various fields like physics [4–6], biology [7–9], and imaging [10–12] among others. Let $\Omega \subset \mathbb{R}^d$, $d = 1, 2, 3$ be an open, bounded set with boundary $\partial\Omega$. Let $\mathfrak{L}$ be a real differential operator between two Banach spaces, which may be either linear or nonlinear. The mathematical formulation of the PDE is given by

$$\mathfrak{L}u = \mathfrak{S}, \tag{1}$$

where u is the unknown quantity to be predicted, and $\mathfrak{S}$ denotes the source term. The form of the operator $\mathfrak{L}$ is typically proposed by experts based on their prior knowledge, observed or measured data, and understanding of the dynamic system. The forward problem, which assumes complete access to the parameters and data defining the problem in Eq. (1), is well-posed in the Hadamard sense [13] and has been extensively studied by mathematicians [14, 15]. However, in practical real-world problems, the parameters and data defining the problem in (1) are often inaccessible or poorly measured. These types of problems, known as inverse problems, require the identification of these parameters or the reconstruction of missing data [16–19]. Unlike forward problems, inverse problems are generally ill-posed, which makes them much more challenging to solve [20–22]. One of the most classical example of inverse problems is the so-called Cauchy problem [23–32]. In this class of problems, both boundary conditions Dirichlet and Neumann are prescribed only on a part of the boundary, while on the remaining part of the boundary called unreachable one, no information is available. Unfortunately, the Cauchy problem is known to be severely ill-posed problems in the sense of Hadamard [13]. Indeed, even if the solution exists and that it is unique, it does not depends continuously on the data. Thus, even a small noises in the given data can cause a large gap between the solution of the original problem and that of the problem with noise data. Hence, a suitable algorithm, which allows to circumvent this ill-posedness phenomenon, is is necessary in order to stably solve the inverse Cauchy problem.

This work investigates a Cauchy problem arising in the context of heat conduction phenomena. In real-world heat conduction problems, two significant challenges often arise: (i) the material's thermal conductivity may exhibit anisotropy, meaning its thermal conductivity varies with direction, requiring careful consideration; and (ii) certain regions of the boundary may be inaccessible for measurements, leading to incomplete thermal boundary conditions. These complexities are typically modeled using the Laplace-Beltrami equation, which describes heat transfer in anisotropic solids. However, the presence of missing boundary conditions introduces significant challenges. To address the lack of boundary data on inaccessible portions of the anisotropic solid's boundary, additional measurements are often employed. These over-specified data, typically collected on the accessible boundaries, are integrated into the model. This approach effectively transforms the problem into an inverse Cauchy problem in the context of anisotropic heat conduction.

As mentioned above, this problem is highly ill-posed. Indeed, even if the solution exists and is unique, it does not depend continuously on the data. Even small amounts of noise in the given data can lead to significant discrepancies between the solution of the original problem and that of the problem with noisy data. Therefore, a suitable algorithm capable of addressing this ill-posedness is crucial for stably solving the inverse Cauchy problem. While much of the literature has focused on solving the inverse Cauchy problem stably in the presence of noise [33–35], noise itself introduces the challenge of scalability. Specifically, for high levels of noise, the additional measurements must be sufficiently abundant to ensure the stability of the solution.

Therefore, this work will discuss the transition from an ill-posed Cauchy problem to a control problem. To do this, we will start by solving two well-posed problems by establishing their weak variational formulation, then by approximating them using the finite element method. By admitting their discrete solutions, we will then approach the resolution of a minimization problem by an iterative method [36]; it is a Limited Memory BFGS, with the aim of ensuring the stability and convergence of the solutions.

2 Data Completion Problem

2.1 Thermal Diffusion Problem: Formulation

Let Ω be a bounded Lipschitz domain of a solid body in $\mathbb{R}^d$, $d = 2$ or 3, with peicewise Lipschitz boundary $\partial\Omega = \Gamma = \Gamma_i \cup \Gamma_a$, such that Γ_a, Γ_i both are not empty and disjoint. The solid in Ω is supposed to be anisotropic, it means that its thermal conductivity varies depending on the direction.

This anisotropy is modeled by a symmetric and positive-definite thermal conductivity tensor $K(x) := [k_{ij}(x)]_{1\leq i,j\leq d}$, and $k_{ij} \in \mathcal{L}^{\infty}(\Omega)$, $i, j = 1, ..., d$.

In the boundary Γ_a, both the Dirichlet and Neumann conditions are considered known or can be measured, it's the accessible boundary, and Γ_i is the inaccessible one, and none of these conditions are known. In the absence of heat sources, the anisotropic heat conduction problem is modeled by the following partial differential equation (PDE)

$$\begin{cases} -\nabla.(K\nabla T) = 0 \; in \; \Omega \\ T = T^* \; in \; \Gamma_a \\ (K\nabla T).n = q^* \; in \; \Gamma_a, \end{cases} \tag{2}$$

where $q(x) = (K(x)\nabla T(x)).n(x)$ is the normal heat flux at the point $x \in \partial\Omega$, T is the temperature distribution in Ω.

2.2 Control Problem

The thermal diffusion problem, defined with boundary conditions on Γ_a and none on Γ_i, is ill-posed in the Hadamard sense due to its sensitivity to input data disturbances. Small changes in boundary measurements can cause large variations in the solution

$T(x)$. This instability is common in inverse problems, where errors in data lead to significant oscillations in the reconstruction. Thus, regularization is necessary to stabilize the solution and reduce noise sensitivity. As a consequence, we reformulate the Cauchy problem into control problem, for that we add a Dirichlet condition (by a boundary control $v \in \mathbb{H}^{\frac{1}{2}}(\Gamma_i)$ on Γ_i) to make it a well-posed direct problem, so instead of looking for the solution on the unstable problem directly, we look for a control function v which influences the solution.

Thus, for each choice of v we obtain a stable solution $T^{(j)}(v)$, $j = 1, 2$, to these well-posed direct problems:

$$\begin{cases} -\nabla.(K\nabla T^{(1)}) = 0 \text{ in } \Omega \\ T^{(1)} = T^* \text{ on } \Gamma_a \\ T^{(1)} = v \text{ on } \Gamma_i, \end{cases} \quad (3) \qquad \begin{cases} -\nabla.(K\nabla T^{(2)}) = 0 \text{ in } \Omega \\ (K\nabla T^{(2)}).n = q^* \text{ on } \Gamma_a \\ T^{(2)} = v \text{ on } \Gamma_i. \end{cases} \quad (4)$$

Finding a solution to the Cauchy problem (2) is equivalent to solving the following control problem:

$$\begin{cases} \text{Find} \quad v \in \mathbb{H}^{\frac{1}{2}}(\Gamma_i), \qquad \text{such that} \\ \quad T^* \in \mathbb{H}^{\frac{1}{2}}(\Gamma_a), \quad q^* \in \mathbb{H}^{-\frac{1}{2}}(\Gamma_a), \quad \text{and} \\ T^{(1)}(v) = T^{(2)}(v). \end{cases} \tag{5}$$

We also define the following functional:

$$\begin{cases} \mathbf{J} : \mathbb{H}^{\frac{1}{2}}(\Gamma_i) \longrightarrow [0, \infty) \\ \qquad\qquad v \longmapsto \frac{1}{2}\|T^{(1)}(v) - T^{(2)}(v)\|_V^2. \end{cases} \tag{6}$$

Where: $T^{(j)}(v) \in \mathbb{H}^1(\Omega)$, $j = 1,\ 2$, was respectively the unique weak solutions of (3) and (4), with $V = \mathbb{L}^2(\Omega)$ or $V = \mathbb{H}^1(\Omega)$. Thus, solving the control problem (5) is equivalent solving the following minimization problem

$$\begin{cases} \text{Find} \quad v^* \in \mathbb{H}^{\frac{1}{2}}(\Gamma_i), \qquad \text{such that} \\ \quad T^* \in \mathbb{H}^{\frac{1}{2}}(\Gamma_a), \qquad q^* \in \mathbb{H}^{-\frac{1}{2}}(\Gamma_a), \quad \text{and} \\ v^* = \arg\min_{v \in \mathbb{H}^{\frac{1}{2}}(\Gamma_i)} \mathbf{J}(v). \end{cases} \tag{7}$$

2.3 Variational Principle

At first, we will examine the variational formulation of the problems to be able to approximate them in finite elements.

We start with the first problem (3). To treat non-homogeneous Dirichlet conditions, we well decompose the solution $T^{(1)}$ on two parts:
– The homogeneous solution $\tilde{T}^{(1)}$.

– The lifting function $R^{(1)}$ which satisfies only the homogeneous boundary conditions.
So the solution will be writing:

$$T^{(1)} = \tilde{T}^{(1)} + R^{(1)}$$

where:

$$\begin{cases} R^{(1)} &= T^* \ on \ \Gamma_a, \\ R^{(1)} &= v \ on \ \Gamma_i. \end{cases}$$

Thus, $\tilde{T}^{(1)}$ is the function which satisfies the homogeneous Dirichlet conditions:

$$\tilde{T}^{(1)} = 0 \ \ in \ \Gamma_a \cup \Gamma_i.$$

So let $\varphi \in \mathbb{H}_0^1(\Omega)$, therefore:

$$\int_\Omega -\nabla.(K\nabla T^{(1)})\varphi d\Omega = 0,$$

according to Green's formula:

$$\int_\Omega -\nabla.(K\nabla T^{(1)})\varphi d\Omega = \int_\Omega (K\nabla T^{(1)})\nabla\varphi d\Omega, \qquad \forall\varphi \in \mathbb{H}_0^1(\Omega).$$

Hence, $\forall\varphi \in \mathbb{H}_0^1(\Omega)$,

$$\int_\Omega (K\nabla\tilde{T}^{(1)})\nabla\varphi d\Omega = -\int_\Omega (K\nabla R^{(1)})\nabla\varphi d\Omega.$$

Then, the variational formulation with lifting associated to the problem (1) is:

$$\begin{cases} \text{Find } \tilde{T}^{(1)} \in \mathbb{H}_0^1(\Omega) \quad \text{such that} \\ \mathcal{A}(\tilde{T}^{(1)}, \varphi) = \mathcal{L}_{R^{(1)}}(\varphi), \quad \forall\varphi \in \mathbb{H}_0^1(\Omega), \end{cases} \tag{8}$$

where:
the bilinear term is: $\mathcal{A}(\tilde{T}^{(1)}, \varphi) = \int_\Omega (K\nabla T^{(1)})\nabla\varphi d\Omega$, and the linear one: $\mathcal{L}_{R^{(1)}}(\varphi) = -\mathcal{A}(R^{(1)}, \varphi) = -\int_\Omega (K\nabla R^{(1)})\nabla\varphi d\Omega$.

For the second problem (4), we also have non-homogeneous Dirichlet conditions, so we will do the same thing, decomposition of the solution $T^{(2)}$, such us

$$T^{(2)} = \tilde{T}^{(2)} + R^{(2)},$$

where:
– $\tilde{T}^{(2)}$ is the homogeneous solution – $R^{(2)}$ is the lifting function, which satisfies only the homogeneous boundary conditions ($R^{(2)} = v \;\; on \;\; \Gamma_i$). So let $\psi \in \mathbb{H}_0^1(\Omega)$, therefore:

$$\int_{\Omega} -\nabla.(K\nabla T^{(2)})\varphi d\Omega \;\; = \;\; 0.$$

The Green's formula will give us:

$$\int_{\Omega}(K\nabla\tilde{T}^{(2)})\nabla\varphi d\Omega = \int_{\Gamma_a} q^*\varphi d\Gamma - \int_{\Omega}(K\nabla R^{(2)})\nabla\varphi d\Omega.$$

So the variational formulation with lifting associated to the problem (2) is:

$$\begin{cases} \text{Find } \tilde{T}^{(2)} \in \mathbb{H}_0^1(\Omega), & \text{such that} \\ \mathcal{A}(\tilde{T}^{(2)}, \varphi) = \mathcal{L}_{R^{(2)}}(\varphi), & \forall\varphi \in \mathbb{H}_0^1(\Omega), \end{cases} \tag{9}$$

where:
the bilinear term is: $\mathcal{A}(\tilde{T}^{(2)}, \varphi) = \displaystyle\int_{\Omega}(K\nabla\tilde{T}^{(2)})\nabla\varphi d\Omega$, and the linear one is: $\mathcal{L}_{R^{(2)}}(\varphi) = \mathcal{L}(\varphi) - \mathcal{A}(R^{(1)}, \varphi)$, such that

$$\mathcal{L}(\varphi) = \int_{\Gamma_a} q^*\varphi d\Gamma, \;\; \text{and} \;\; \mathcal{A}(R^{(2)}, \varphi) = \int_{\Omega}(K\nabla R^{(2)})\nabla\varphi d\Omega.$$

2.4 Finite Element Approximation

Let consider a regular family of meshes $(\tau_h)_h$ on Ω, composed of closed triangles. The discretization parameter h signifies the maximal diameter of the elements of τ_h. Let $(V_h)_h \subset \mathbb{H}_0^1(\Omega)$ a family of finite-dimensional vector spaces constructed according to τ_h.
So the discrete variational problem of (8) is given by:

$$(DVP1)\begin{cases} \text{Find } \tilde{T}_h^{(1)} \in V_h \quad \text{such that} \\ \displaystyle\int_{\Omega}(K\nabla\tilde{T}_h^{(1)})\nabla\varphi_h d\Omega = \; - \int_{\Omega}(K\nabla R_h^{(1)})\nabla\varphi_h d\Omega, \forall\varphi_h \in V_h. \end{cases}$$

For (9) the discrete problem is given by:

$$(DVP2)\begin{cases} \text{Find } \tilde{T}_h^{(2)} \in V_h \quad \text{such that} \\ \displaystyle\int_{\Omega}(K\nabla\tilde{T}_h^{(2)})\nabla\varphi_h d\Omega = \int_{\Gamma_a} q^*\varphi_h d\Gamma - \int_{\Omega}(K\nabla R_h^{(2)})\nabla\varphi_h d\Omega, \forall\varphi_h \in V_h. \end{cases}$$

Let $(\psi_i)_{i=1,...,N}$ be a basis of the vector space V_h, such us

$$\tilde{T}_h^{(1)}(x) = \sum_{i=1}^{N} \tilde{T}_i^{(1)} \psi_i(x), \quad \tilde{T}_h^{(2)}(x) = \sum_{i=1}^{N} \tilde{T}_i^{(2)} \psi_i(x),$$

we choose the test functions as follows

$$\varphi_h(x) = \psi_j(x), \quad for \quad j = 1, ..., N.$$

So

$$(DVP1) \Leftrightarrow \begin{cases} \text{Find } \tilde{T}_h^{(1)} \in V_h \quad \text{such that} \\ \int_\Omega (K\nabla \tilde{T}_h^{(1)})\nabla\psi_j d\Omega = -\int_\Omega (K\nabla R_h^{(1)})\nabla\psi_j d\Omega, \quad \forall j = 1, ..., N \end{cases}$$

$$\Leftrightarrow \begin{cases} \text{Find } \tilde{\mathbf{T}}^{(1)} \in \mathbb{R}^N \quad \text{such that} \\ \sum_{i=1}^{N} \int_\Omega K\nabla\psi_i \nabla\psi_j \tilde{T}_i^{(1)} d\Omega = -\int_\Omega (K\nabla R_h^{(1)})\nabla\psi_j d\Omega, \quad \forall j = 1, ..., N, \end{cases}$$

hence, the linear system is given by

$$\mathbf{A}\tilde{\mathbf{T}}^{(1)} = F^{(1)}, \tag{10}$$

where:

$$\mathbf{A} = (A_{ij})_{1\leq i,j\leq N} = \left(\int_\Omega K\nabla\psi_i\nabla\psi_j d\Omega\right)_{1\leq i,j\leq N},$$

end

$$F = (F_i)_{1\leq i\leq N} = \left(-\int_\Omega (K\nabla R_h^{(1)})\nabla\psi_j d\Omega\right)_{1\leq i\leq N}.$$

For the second problem, we do the same thing, so

$$(DVP2) \Leftrightarrow \begin{cases} \text{Find } \tilde{T}_h^{(2)} \in V_h \quad \text{such that} \\ \int_\Omega (K\nabla \tilde{T}_h^{(2)})\nabla\psi_j d\Omega = -\int_\Omega (K\nabla R_h^{(2)})\nabla\psi_j d\Omega, \forall j = 1, ..., N \end{cases}$$

$$\Leftrightarrow \begin{cases} \text{Find } \tilde{\mathbf{T}}^{(2)} \in \mathbb{R}^N \quad \text{such that} \\ \sum_{i=1}^{N} \int_\Omega K\nabla\psi_i \nabla\psi_j \tilde{T}_i^{(2)} d\Omega = \int_{\Gamma_a} q^* \psi_i d\Gamma - \int_\Omega (K\nabla R_h^{(2)})\nabla\psi_j d\Omega, \quad \forall j = 1, ..., N. \end{cases}$$

So, the according linear system is given by

$$\mathbf{A}\tilde{\mathbf{T}}^{(2)} = F^{(2)}, \tag{11}$$

where

$$F^{(2)} = (F_i)_{1\leq i\leq N} = \left(\int_{\Gamma_a} q^*\psi_i d\Gamma - \int_\Omega (K\nabla R_h^{(2)})\nabla\psi_i d\Omega\right)_{1\leq i\leq N}.$$

So from (10) and (11):

$$\begin{cases} \tilde{\mathbf{T}}^{(1)} = \mathbf{A}^{-1}F^{(1)} \\ \tilde{\mathbf{T}}^{(2)} = \mathbf{A}^{-1}F^{(2)}. \end{cases} \tag{12}$$

where A is invertible, since K is symmetric and positive-definite.

Now, we introduce the minimization problem (7) in the discrete form:

$$\begin{aligned} \mathbf{J}_h(\mathbf{v}) &= \tfrac{1}{2}\|\tilde{\mathbf{T}}^{(1)}(\mathbf{v}) - \tilde{\mathbf{T}}^{(2)}(\mathbf{v})\|_V^2 \\ &= \tfrac{1}{2}\|\mathbf{A}^{-1}F^{(1)}(\mathbf{v}) - \mathbf{A}^{-1}F^{(2)}(\mathbf{v})\|_V^2 \\ &= \tfrac{1}{2}\|\mathbf{A}^{-1}(F^{(1)} - F^{(2)})(\mathbf{v})\|_V^2. \end{aligned} \tag{13}$$

So the minimization problem is given by:

$$\begin{cases} \text{Find} \quad \mathbf{v}^* \in \mathbb{R}^N, \quad \text{such that} \\ B = F^{(1)} - F^{(2)} \in \mathbb{R}^N, \quad \text{and} \\ \mathbf{v}^* = \arg\min_{\mathbf{v}\in\mathbb{R}^N} \mathbf{J}_h(v). \end{cases} \tag{14}$$

3 Limited Memory BFGS

The L-BFGS (Limited-memory Broyden-Fletcher-Goldfarb-Shanno) is a popular optimization algorithm for high-dimensional problems that do not allow the storage of full Hessian. L-BFGS works with a small amount of memory so that the inverse Hessian matrix does not need to be specified, although it can be approximated. It's an iterative method similar to BFGS, based on the quasi-Newton approach, which is, however, designed for large-scale optimization. It is efficient in terms of memory storage as it only saves a few steps back of the gradients and is thus ideal for high-dimensional applications.

The L-BFGS algorithm should be chosen to solve high-dimensional optimization problems, as in the present case (14) of a difference minimization problem between two discretized solutions, $\tilde{\mathbf{T}}^{(1)}(\mathbf{v})$ and $\tilde{\mathbf{T}}^{(2)}(\mathbf{v})$. When the problem is located in high-dimensional space, the standard BFGS algorithm has to overcome memory limitations due to its requirement to store the full Hessian matrix or its inverse. Instead, L-BFGS uses a limited-memory approach that stores a few gradients of the past iterations, which is much more memory efficient.

3.1 Algorithm and Results

In our algorithm, L-BFGS is used to find the optimal vector v^* by minimizing the objective function $\mathbf{J}_h$. It will be achieved without specifying the inverse of the Hessian matrix of $\mathbf{J}_h$, which is approximated.

Due to its iterative nature and its ability to efficiently handle historical gradient information, the L-BFGS algorithm is particularly well suited to high-dimensional applications. It only keeps in memory the gradient information needed for the current step, which decreases storage overhead while speeding up convergence, especially when solving complex linear systems at each iteration. The attached algorithm applies this method by minimizing $\mathbf{J}_h$.

Algorithm 1 Minimization of $\mathbf{J}_h(v) = \frac{1}{2}\|\tilde{T}^{(1)}(v) - \tilde{T}^{(2)}(v)\|_V^2$ using L-BFGS

Require: Initial vector v_0, maximum number of iterations `maxiter`
Ensure: Solution v^* that minimizes $\mathbf{J}_h(v)$

Initialize $v \Leftarrow v_0$
for iteration $k = 1, 2, \ldots,$ `maxiter` **do**
  Step 1: Solve the system to obtain $\tilde{T}^{(1)}(v)$
    Solve problem (1) using L-BFGS $A\tilde{T}^{(1)}(v) = F^{(1)}(v)$
  Step 2: Update $v \Leftarrow v + \alpha d_k$ **using a search step** α
  Step 3: Solve the system to obtain $\tilde{T}^{(2)}(v)$
    Solve problem (2) using L-BFGS $A\tilde{T}^{(2)}(v) = F^{(2)}(v)$
  Step 4: Compute the objective function and gradient
    Compute $\mathbf{J}_h(v) = \frac{1}{2}\|\tilde{T}^{(1)}(v) - \tilde{T}^{(2)}(v)\|_V^2$
    Compute the gradient $\nabla\mathbf{J}_h(v)$ (approximated or analytical)
end for
return $v^* = v$

Example 1 We consider $\Omega =]0,1[\times]0,b[$ where $b > 0$. We shall solve the following Cauchy problem for the anisotropic heat conduction problem

$$\begin{cases} \nabla(K\nabla T) & = 0, & \text{in} &]0,1[\times]0,b[, \\ T(x,0) & = f_1, & \text{on} & 0 \le x \le 1, \\ \partial_y T(x,0) & = f_2, & \text{on} & 0 \le x \le 1, \\ T(0,y) = T(1,y) & = 0, & \text{on} & 0 \le y \le b. \end{cases} \tag{15}$$

We take $\Gamma_a =]0,1[\times\{0\}$ and $\Gamma_i =]0,1[\times\{b\}$. for the numerical computations, we particularly choose $b = 0.2$, $N = 400$, $M = 80$, and select the boundary data $f_1(x) = T(x,0)$ on Γ_a as

$$T(x,0) = \left(3\sin(\pi x) + \frac{\sin(3\pi x)}{19} + 9\exp(-30(x-b)^2)\right)x^2(1-x)^2.$$

The exact boundary data on Γ_i, used to test the performance of the algorithm is given by

$$Te(x,b) = 2\left(8\sin(\pi x) + \frac{\sin(3\pi x)}{17} + 20\exp(-50(x-b)^2)\right)x^2(1-x)^2.$$

Note that the exact solution T is unknown, then first we construct it by solving the direct problem for the anisotropic heat conduction problem 15. The graph illustrates the efficiency of the L-BFGS method by comparing the reconstructed results with the exact solution. We observe that the two curves agree, so the algorithm has converged well, and the reconstructed solution is close to the exact solution.

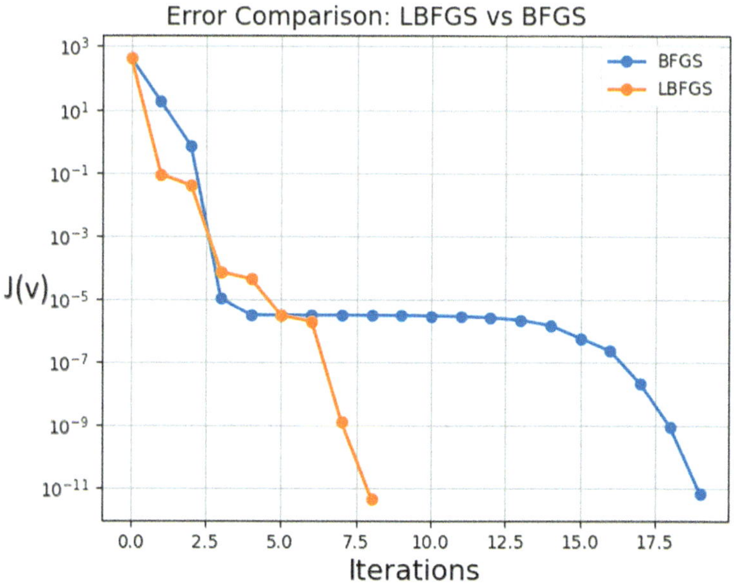

In the figure above, we can see that LBFGS (the orange graph) converges very quickly, reaching extremely low errors 10^{-11} in around 7 iterations, while BFGS (the blue one) is slower, and although it reduces $J(v)$, it takes a long time to reach similar values. This proves the effectiveness of LBFGS (Fig. 1).

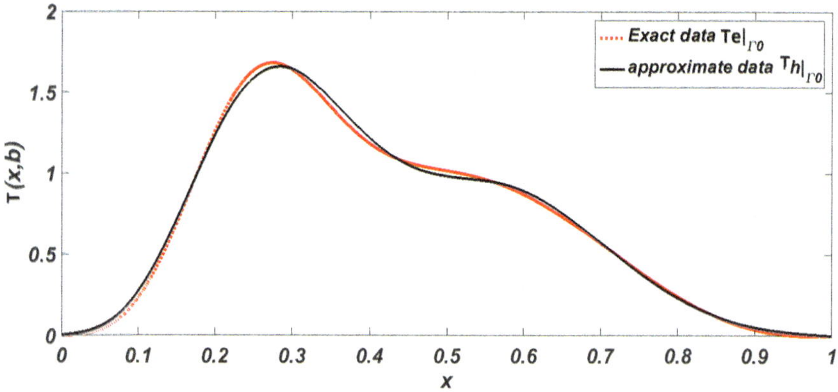

Fig. 1 Exact solution at $y = b$ and the reconstructed one, obtained by the algorithm LBFGS

4 Conclusion

In summary, this work addresses the Cauchy problem in anisotropic heat conduction, a classical example of an ill-posed inverse problem. We have proposed a rigorous methodology that reformulates the problem as a control problem with regularization, ensuring the stability and convergence of solutions. The chosen approach utilizes finite element method approximation and iterative optimization via the limited-memory BFGS method.

The results demonstrate that this method is effective in reconstructing inaccessible data with remarkable accuracy, as evidenced by the low errors achieved. This contribution represents a significant advancement in the field of regularization and inverse problem-solving for anisotropic heat conduction. Finally, this approach paves the way for future applications where the stability and convergence of solutions are crucial.

References

1. Laghrib, A., Afraites, L.: Image denoising based on a variable spatially exponent PDE. Appl. Comput. Harmonic Anal. **68**, 101608 (2024)
2. Laghrib, A.: On some evolution equation with combined local and nonlocal p $(x,[\nabla u])$-Laplace operator for image denoising. J. Franklin Insti. **361**(5), 106667 (2024)
3. Nachaoui, M., Afraites, L., Laghrib, A.: A regularization by denoising super-resolution method based on genetic algorithms. Signal Process. Image Commun. **99**, 116505 (2021)
4. Chakib, A., Hadri, A., Nachaoui, A., Nachaoui, M.: Homogenization of parabolic problem with nonlinear transmission condition. Nonlinear Anal. Real World Appl. **37**, 433–456 (2017)
5. Jauberteau, F., Nachaoui, M., Zaroual, S.: A novel identification scheme of an inverse source problem based on Hilbert reproducing kernels. In: New Trends of Mathematical Inverse Problems and Applications, pp. 157–175. Springer, Cham (2023)
6. Oulmelk, A., Srati, M., Afraites, L.: Comparing numerical methods for inverse source problem in time-fractional diffusion equation. In: New Trends of Mathematical Inverse Problems and Applications, pp. 1–16. Springer, Cham (2023)
7. Akrid, T., Elazzouzi, A., Ouhinou, A.: Asymptotic almost periodicity of bounded solutions for some partial functional differential equations with delay. Discrete Continuous Dyn. Syst. S **18**(1), 321–349 (2025)
8. Ladib, M., Mambili-Mamboundou, H., Ouhinou, A.: Mathematical analysis of the spatiotemporal concentration of growth inhibitory factor in multicellular spheroids tumor. Discrete Continuous Dyn. Syst.-S **18**(1), 182–196 (2025)
9. Nachaoui, A., Nachaoui, M., Tadumadze, T.: Meshless methods to noninvasively calculate neurocortical potentials from potentials measured at the scalp surface. In: New Trends of Mathematical Inverse Problems and Applications, pp. 99–117. Springer, Cham (2023)
10. Nachaoui, A., Laghrib, A., Hadri, A., Nachaoui, M.: A novel variable exponent PDE with dependency on $\gamma(u, |\nabla u0, \sigma|)$ for image despeckling application. Nonlinear Anal. Real World Appl. **83**, 104264 (2025)
11. Nachaoui, M., Laghrib, A., Hakim, M.: A new space-variant optimization approach for image segmentation. In: Mathematical Control and Numerical Applications, pp. 87–97. Springer, Cham (2021)

12. El Hakoume, A., Zaabouli, Z., Laghrib, A., Afaites, L.: Image restoration using a coupled reaction-diffusion equations. In: New Trends of Mathematical Inverse Problems and Applications, pp. 137–155. Springer, Cham (2023)
13. Hadamard, J.: Lectures on Cauchy's Problem in Linear Partial Differential Equations, p. 316. Dover Publications, New York (1953)
14. Progress in Partial Differential Equations: Springer Proceedings in Mathematics & Statistics, vol. 44, p. 447. Springer, Germany (2013)
15. Numerical solution of partial differential equations: theory, algorithms, and their applications. Springer Proc. Math. Stat. **45**, 327 (2013). https://doi.org/10.1007/978-1-4614-7172-1
16. Isakov, V.: Inverse Problems for Partial Differential Equations, vol. 127, p. 406. Springer, Germany (2006). https://doi.org/10.1007/978-3-319-51658-5
17. Banks, H.T., Kunisch, K.: Estimation Techniques for Distributed Parameter Systems. Springer, Germany (2012)
18. Kabanikhin, S.I.: Inverse and Ill-Posed Problems: Theory and Applications. de Gruyter, Germany (2011)
19. Nachaoui, A., Aboud, F.: Solving geometric inverse problems with a polynomial based meshless method. In: New Trends of Mathematical Inverse Problems and Applications, pp. 119–136. Springer, Cham (2023)
20. Chakib, A., Nachaoui, A., Nachaoui, M.: Existence analysis of an optimal shape design problem with non coercive state equation. Nonlinear Anal. Real World Appl. **28**, 171–183 (2016)
21. Hadri, A., Nachaoui, M., Laghrib, A., Chakib, A., Afraites, L.: A primal-dual approach for the robin inverse problem in a nonlinear elliptic equation: the case of the $l_1 - l_2$ cost functional. J. Inverse Ill-Posed Probl. **31**(2), 243–266 (2023)
22. Chakib, A., Nachaoui, A., Nachaoui, M.: Finite element approximation of an optimal design problem. J. Appl. Comput. Math. **11**(1), 19–26 (2012)
23. Bergam, A., Chakib, A., Nachaoui, A., Nachaoui, M.: Adaptive mesh techniques based on a posteriori error estimates for an inverse Cauchy problem. Appl. Math. Comput. **346**, 865–878 (2019)
24. Chakib, A., Nachaoui, A., Nachaoui, M., Ouaissa, H.: On a fixed point study of an inverse problem governed by stokes equation. Inverse Prob. **35**(1), 015008 (2018)
25. Jourhmane, M., Nachaoui, A.: A relaxation algorithm for solving a Cauchy problem. In: On inverse problems in engineering, engineering. In: Proceedings of the 2nd International Conferences, vol. 1, pp. 151–158 (1996)
26. Nachaoui, A.: An improved implementation of an iterative method in boundary identification problems. Numer. Algorithms **33**(1–4), 381–398 (2003)
27. Nachaoui, A., Nachaoui, M., Chakib, A., Hilal, M.A.: Some novel numerical techniques for an inverse Cauchy problem. J. Comput. Appl. Math. **381**, 113030 (2021)
28. Cheng, H., Zhu, P., Gao, J.: A regularization method for the Cauchy problem of the modified Helmholtz equation. Math. Methods Appl. Sci. **38**(17), 3711–3719 (2015)
29. Mihai Bucataru, L.M. Iulian Cîmpean: a gradient-based regularization algorithm for the Cauchy problem in steady-state anisotropic heat conduction. Comput. Math. Appl. **119**:220–240 (2022)
30. Liu, C.-S., Wang, F.: A meshless method for solving the nonlinear inverse Cauchy problem of elliptic type equation in a doubly-connected domain. Comput. Math. Appl. **76**(8), 1837–1852 (2018)
31. Nachaoui, A., Rashid, S.M.: A mesh free wavelet method to solve the Cauchy problem for the Helmholtz equation. In: The International Conference on New Trends in Applied Mathematics, pp. 77–98 (2022)
32. Wang, F., Fan, C.-M., Hua, Q., Gu, Y.: Localized MFS for the inverse Cauchy problems of two-dimensional Laplace and biharmonic equations. Appl. Math. Comput. **364**, 124658–14 (2020)
33. Hadri, A., Laghrib, A., Nachaoui, A., Nachaoui, M.: On the resolution of the non-smooth inverse Cauchy problem by the primal-dual method. Numer. Algorithms 1–34 (2024)
34. Berdawood, K.A., Nachaoui, A., Nachaoui, M.: An accelerated alternating iterative algorithm for data completion problems connected with Helmholtz equation. Stat. Optim. Inf. Comput. **11**(1), 2–21 (2023)

35. Nachaoui, A., Aboud, F., Nachaoui, M.: Acceleration of the KMF algorithm convergence to solve the Cauchy problem for Poisson's equation. In: Mathematical Control and Numerical Applications: JANO13, Khouribga, Morocco, pp. 43–57 (2021)
36. Liu, D.C., Nocedal, J. (eds.): On the limited memory BFGS method for large scale optimization. Math. program. **45** (1989)

Identification of a Space Dependent Force in a Hyperbolic Equation Using a Polynomial Expansion

Abdeljalil Nachaoui and Gulnar W. Sadiq

Abstract This paper presents a detailed study on the application of a polynomial expansion method to solve inverse problems related to the determination of the **displacement** and **space-dependent forces** on a vibrating structure based on observed final displacement data. Beginning with a structured problem formulation, we analyze the challenges of ill-conditioning and solution instability often encountered in the obtained discret systems. To address these issues, we introduce a regularized approach, transforming the original system to enhance numerical stability. Additionally, a preconditioning technique is employed to further improve convergence and computational efficiency. We demonstrate the effectiveness of this method through numerical experiments, analyzing the impact of both regularization and preconditioning on solution accuracy.

Keywords Inverse force problem · Polynomial expansion · Wave equation · CGLS method

1 Introduction

This work addresses the challenge of force identification from measured data in the context of the hyperbolic wave equation. Solving this inverse problem is essential for numerous applications involving unknown force distributions and control requirements. Due to the partial unknown nature of the forces driving the physical phenomenon, it is necessary to measure specific effects that allow compensating for this missing information. Selecting an appropriate measurement is crucial for formulating a successful inverse problem, where a suitable choice ensures that the

A. Nachaoui (✉)
Laboratoire de Mathématiques Jean Leray, Nantes Université, Nantes, France
e-mail: Abdeljalil.Nachaoui@univ-nantes.fr

G. W. Sadiq
Mathematical Sciences Department, College of Basic Education, University of Sulaimani, Sulaimani, Iraq
e-mail: gulnar.sadiq@univsul.edu.iq

A. Laghrib et al. (eds.), *Computational Methods for Inverse Problems and Applications*, Springer Proceedings in Mathematics & Statistics 498,
https://doi.org/10.1007/978-3-031-89498-5_9

unknown force can be uniquely determined from additional observations. However, stability remains generally unattainable.

This problem is part of a broader class of inverse problems [1–15], attracting substantial interest across physics, engineering, medicine, and environmental science [16–22]. Such problems are known to be ill-posed [23–25], necessitating methods that handle this characteristic appropriately.

Prior work has focused on reconstructing space-dependent forces in the wave equation using Cauchy data, specifically displacement and its normal derivative on the boundary [26]. Although solution uniqueness holds [9, 27, 28], this inverse problem is inherently ill-posed due to the discrepancy between the unknown force $f(x)$ defined over Ω and the known input data, such as u and ∂_ν, which vary on $\partial\Omega \times (0, T)$. Consequently, measuring the displacement $u(x, T)$ over the domain at a specific time $t = T$ appears more natural, aligning the directions of the overposed data and the unknown force [29]. Space-dependent force identification from this upper-bound displacement measurement $u(x, T)$ has been theoretically validated for uniqueness [30]. Additional studies have also used final-time displacement data to address related inverse problems, leveraging weak solution approaches or quasi-nonlinearity to address non-uniqueness challenges [31, 32].

A related inverse problem arises in heat transfer, where time-averaged displacement measurements are used for force identification within the parabolic heat equation [33].

Inverse problem-solving techniques can be divided into two main categories: iterative and direct methods [2–4, 9, 12, 34–45]. Iterative methods, which include those previously mentioned and the conjugate gradient technique [46], typically involve initializing with an estimate and refining it through repeated minimization or solution of well-posed sub-problems. These methods, while often computationally intensive, have recently seen enhancements aimed at reducing iteration counts [47, 48]. Nevertheless, convergence is not achievable in a single iteration.

Direct methods, conversely, are more computationally efficient, often requiring only one solver. Classic examples include the quasi-reversibility method [20, 49, 50], which reduces computation time but may suffer from numerical instability.

These methods necessitate the discretization of the boundary problem, often achieved using finite difference (FDM), finite element (FEM), or boundary element methods (BEM). Challenges associated with these methods stem from domain meshing requirements. Recently, meshless methods, which replace meshing with a set of arbitrarily distributed nodes, have been proposed [10, 41, 45, 51–58].

Among meshless techniques, polynomial expansion methods stand out [6, 15, 16, 54, 56, 57, 59], providing flexible and robust algorithms suitable for most boundary value problems (BVPs). Here, we combine polynomial expansion with multi-scale preconditioning for enhanced performance [16, 33, 56].

The remainder of this paper is organized as follows: Sect. 2 introduces the problem, Sect. 3 presents the solution methodology, Sect. 4 provides examples demonstrating the method's efficiency, and the final section concludes with remarks.

2 Problem Formulation

Consider a medium represented by the interval $\Omega = (0, L)$, where $L > 0$ is the length of a vibrating string. We define the space-time cylinder $Q_T = \{(x, t) \mid 0 < x < L,\ 0 < t < T\}$, where $T > 0$ denotes the observation time. Our goal is to determine the **displacement** $u(x, t)$ and the **force** $f(x)$ in the context of the hyperbolic wave equation.

$$\frac{\partial^2 u(x,t)}{\partial t^2} - \frac{\partial^2 u(x,t)}{\partial x^2} = f(x)g(x,t) + \chi(x,t), \quad (x,t) \in Q_T, \tag{1}$$

where $g(x, t)$ and $\chi(x, t)$ are known functions.

The solution to Eq. (1) must satisfy the following initial conditions:

$$u(x,0) = \phi(x), \quad u_t(x,0) = \psi(x), \quad x \in \Omega, \tag{2}$$

the Dirichlet boundary conditions:

$$u(0,t) = w_0(t), \quad u(L,t) = w_1(t), \quad t \in (0,T), \tag{3}$$

and the final displacement measurement:

$$u(x,T) = U_T(x), \quad x \in \Omega. \tag{4}$$

3 Methodology

To approximate the displacement $u(x, t)$ and the force $f(x)$ in the hyperbolic wave equation for the problem (1)–(4), we employ a polynomial expansion method combined with the collocation technique. The key idea [33] is to expand $\frac{\partial^4 u}{\partial t^2 \partial x^2}$ as a polynomial in x and t, as follows:

$$\frac{\partial^4 u}{\partial t^2 \partial x^2} = \sum_{i=1}^{m_u} \sum_{j=1}^{i} c_{ij}^u x^{i-j} t^{j-1}, \tag{5}$$

where m_u is the polynomial degree and c_{ij}^u are the unknown coefficients to be determined, with $n = \frac{m_u(m_u+1)}{2}$ coefficients in total. We then integrate (5) and use the conditions (2)–(4) to obtain an analytical solution for $u(x, t)$.

Suppose further that the expression for $f(x)$ can be represented as:

$$f(x) = \sum_{k=1}^{m_f} c_k^f x^{k-1}. \tag{6}$$

These expressions are substituted into Eq. (1), and the resulting equation is evaluated at collocation points within the domain Q_T, yielding a system of linear equations with $n + m_f$ unknown coefficients c_{ij}^u and c_k^f.

To find a pair of functions $(f(x), u(x,t))$ that satisfies the problem (1)–(4) on Q_T, we integrate (5) twice with respect to t from 0 to t:

$$\frac{\partial^2 u(x,t)}{\partial x^2} = \sum_{i=1}^{m_u}\sum_{j=1}^{i} c_{ij}^u \frac{x^{i-j}t^{j+1}}{j(j+1)} + \frac{\partial^2 u(x,0)}{\partial x^2} + \frac{\partial^3 u(x,0)}{\partial t \partial x^2} t. \tag{7}$$

From condition (2), we know $u(x,0) = \phi(x)$ and $\frac{\partial u(x,0)}{\partial t} = \psi(x)$. Thus, $\frac{\partial^2 u(x,0)}{\partial x^2} = \phi''(x)$ and $\frac{\partial^3 u(x,0)}{\partial t \partial x^2} = \psi''(x)$. Substituting these values into (7) gives:

$$\frac{\partial^2 u(x,t)}{\partial x^2} = \sum_{i=1}^{m_u}\sum_{j=1}^{i} c_{ij}^u \frac{x^{i-j}t^{j+1}}{j(j+1)} + \phi''(x) + \psi''(x)t. \tag{8}$$

Integrating (8) twice with respect to x from 0 to x results in:

$$\begin{aligned} u(x,t) = {} & \sum_{i=1}^{m_u}\sum_{j=1}^{i} c_{ij}^u \frac{t^{j+1}x^{i-j+2}}{j(j+1)(i-j+1)(i-j+2)} + \phi(x) - \phi(0) - x\phi'(0) \\ & + t\psi(x) - t\psi(0) - xt\psi'(0) + x\frac{\partial u(0,t)}{\partial x} + u(0,t). \end{aligned} \tag{9}$$

Setting $x = L$ in (9) and using the boundary conditions (3) yields:

$$\begin{aligned} w_1(t) = {} & \sum_{i=1}^{m_u}\sum_{j=1}^{i} c_{ij}^u \frac{t^{j+1}L^{i-j+2}}{j(j+1)(i-j+1)(i-j+2)} + \phi(L) - \phi(0) - L\phi'(0) \\ & + t\psi(L) - t\psi(0) - Lt\psi'(0) + L\frac{\partial u(0,t)}{\partial x} + w_0(t). \end{aligned} \tag{10}$$

Then,

$$\begin{aligned} \frac{\partial u(0,t)}{\partial x} = {} & -\sum_{i=1}^{m_u}\sum_{j=1}^{i} c_{ij}^u \frac{t^{j+1}L^{i-j+2}}{j(j+1)(i-j+1)(i-j+2)} \\ & + \frac{1}{L}\{w_1(t) - w_0(t) - \phi(L) + \phi(0) - t\psi(L) + t\psi(0)\} + \phi'(0) + t\psi'(0). \end{aligned} \tag{11}$$

Substituting (11) into (9) gives:

$$u(x,t) = \sum_{i=1}^{m_u}\sum_{j=1}^{i} \frac{t^{j+1}}{j(j+1)(i-j+1)(i-j+2)}\left[x^{i-j+2} - xL^{i-j+1}\right]c^u_{ij} + \phi(x) - \phi(0)$$
$$+t\psi(x) - t\psi(0) - \frac{x}{L}(\phi(L) - \phi(0) + t\psi(L) - t\psi(0)) + \frac{x}{L}w_1(t) + (1-\frac{x}{L})w_0(t). \quad (12)$$

Differentiating (12) twice with respect to t gives:

$$\frac{\partial^2 u(x,t)}{\partial t^2} = \sum_{i=1}^{m_u}\sum_{j=1}^{i} \frac{t^{j-1}}{(i-j+1)(i-j+2)}\left[x^{i-j+2} - xL^{i-j+1}\right]c^u_{ij} + \frac{x}{L}w_1''(t) + (1-\frac{x}{L})w_0''(t) \quad (13)$$

Substituting (8) and (13) into (1) yields:

$$\sum_{i=1}^{m_u}\sum_{j=1}^{i}\left[\frac{t^{j-1}(x^{i-j+2} - xL^{i-j+1})}{(i-j+1)(i-j+2)} - \frac{x^{i-j}t^{j+1}}{j(j+1)}\right]c^u_{ij} - f(x)g(x,t)$$
$$= \chi(x,t) + \phi''(x) + t\psi''(x) - \frac{x}{L}w_1''(t) - (1-\frac{x}{L})w_0''(t). \quad (14)$$

Setting $t = T$ in Eq. (12) and using condition (4) gives:

$$\sum_{i=1}^{m_u}\sum_{j=1}^{i} \frac{T^{j+1}}{j(j+1)(i-j+1)(i-j+2)}\left[x^{i-j+2} - xL^{i-j+1}\right]c^u_{ij}$$
$$= U_T(x) - \phi(x) + \phi(0) + \frac{x}{L}[\phi(L) - \phi(0)] + \frac{x}{L}[T\psi(L) - T\psi(0)] - T\psi(x) + T\psi(0)$$
$$-\frac{x}{L}w_1(T) - (1-\frac{x}{L})w_0(T). \quad (15)$$

Evaluating Eq. (14) at n_{int} collocation points in Q_T and Eq. (15) at n_b points in $(0, L)$ yields a linear system with $n = \frac{m_u(m_u+1)}{2} + m_f$ unknowns. The overdetermined linear system is given by:

$$A\mathbf{C} = B, \quad (16)$$

where B is an n_{total}-dimensional vector, A is an $n_{\text{total}} \times n$ matrix, and $\mathbf{C}$ is the reordered vector of unknowns (Fig. 1).

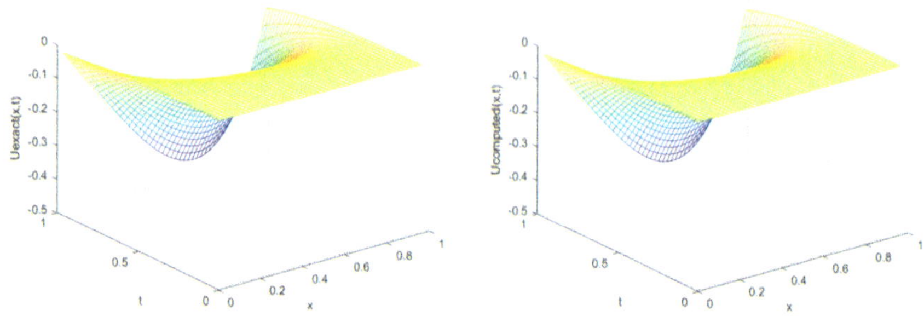

(a) Exact Solution (b) Computed Solution

Fig. 1 Example 1: Exact and computed displacement without regularization or preconditioning

4 Solving the Linear System

To solve the over-determined linear system (16) and address potential ill-conditioning issues with matrix A, we reformulate the problem as a regularized least-squares minimization:

$$\min_{\mathbf{C}\in\mathbb{R}^n} \left\{ \|A\mathbf{C} - b\|^2 + \mu^2 \|\mathbf{C}\|^2 \right\}, \tag{17}$$

where $\mu > 0$ is a regularization parameter. Solving the minimization in (17) is equivalent to solving the regularized linear system:

$$\left[A^T A + \mu I\right] \mathbf{C}_\mu = A^T b. \tag{18}$$

Setting $\mu = 0$ in this system yields the normal equation:

$$D\mathbf{C} = b_1, \tag{19}$$

where $b_1 = A^T b$ and $D = A^T A$ is a symmetric positive definite matrix, and where $\mathbf{C}$ is the original solution of (16). As μ approaches zero, the solution to the regularized system (18) can approximate the solution of the original system (16).

In cases where further improvement in solution accuracy or computational efficiency is needed, we employ preconditioning techniques. Specifically, we apply two-sided k-multi-preconditioning (TSkMP) as outlined in [60], or alternatively, we may select other preconditioning methods described in [61, 62] and their references.

5 Numerical Results

This section presents the numerical results from our study, focusing on the effectiveness of the polynomial expansion method, the influence of regularization and preconditioning on solving linear systems, and the stability of the solutions when noise is introduced into the data. Figures and tables are included to visually demonstrate the outcomes and to enhance the understanding of our findings (Figs. 2 and 3).

Example 1 Consider the case where both $u(x,t)$ and $f(x)$ are polynomials. Given $u_{\text{exact}} = x^4t^3 - xt^3$ and $f_{\text{exact}} = x^2 + x + 1$, with domain $Q_T = \{(x,t) \mid 0 < x < 1,\ 0 < t < 1\}$, additional parameters include:

$$g(x,t) = x^4t^2 - 12x^2t^3, \quad \chi(x,t) = 6x^4t - 6xt - x^6t^2 + 12x^4t^3 - x^5t^2 + 12x^3t^3 - x^4t^2,$$

$$\phi(x) = 0, \quad \psi(x) = 0, \quad w_0(t) = 0, \quad w_1(t) = 0, \quad U_T(x) = x^4 - x.$$

To find suitable values for parameters mf and μ, an inverse analysis was conducted by varying mf and μ within the range (1–5), using the $CGLS$ algorithm with different tolerances and different numbers of interior points N_x and N_t. An optimal parameter set that yields accurate approximations of errors D and Z, defined as

$$D = \|U_{\text{exact}}(x,t) - U_{\text{computed}}(x,t)\|_\infty, \quad Z = \|f_{\text{exact}}(x,t) - f_{\text{computed}}(x,t)\|_\infty,$$

is given below: Parameters: $\mu = 5$, $mf = 3$, $N_x = 50$, $N_t = 50$. Number of iterations: 81, with tol $= 10^{-15}$ (Fig. 4).

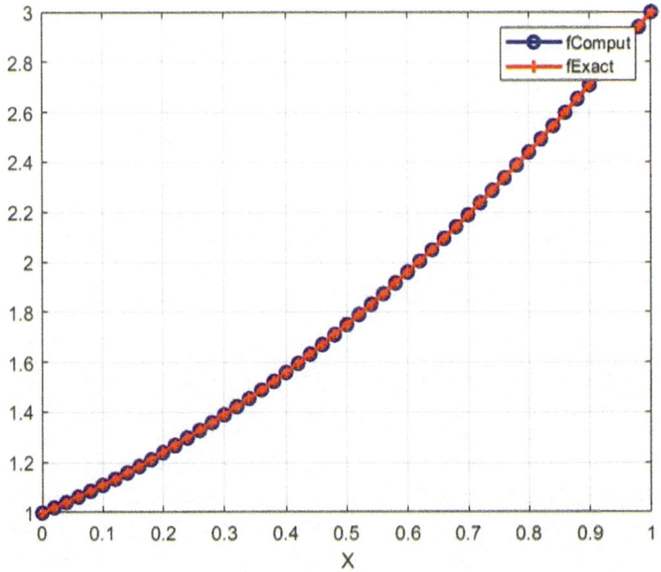

Fig. 2 Exact and computed force $f(x)$ for Example 1, without regularization or preconditioning

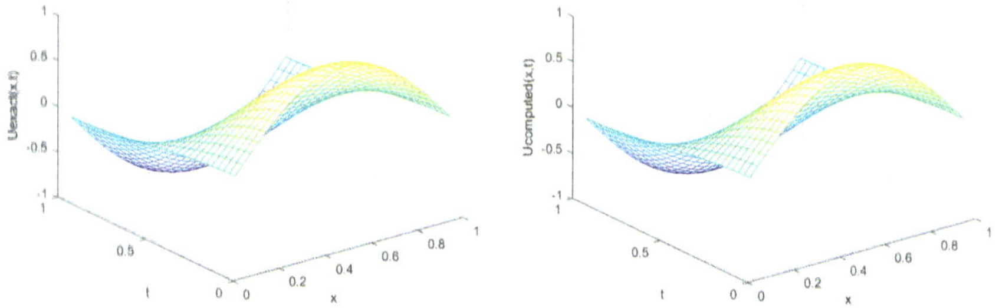

(a) Exact Solution (b) Computed Solution

Fig. 3 Example 2: Exact and computed displacement without regularization or preconditioning

Example 2 Consider the case where both $u(x,t)$ and $f(x)$ are non-polynomials. Let $u_{\text{exact}} = \sin(\pi x)\cos(\pi t)$ and $f(x) = \frac{1}{\sqrt{2\pi}}\exp\left(-\frac{(x-0.5)^2}{2}\right)$, with domain $Q_T = \{(x,t) \mid 0 < x < 1,\ 0 < t < 1\}$, and additional parameters:

$$g(x,t) = 1 + t, \quad \chi(x,t) = -(1+t)\left(\frac{1}{\sqrt{2\pi}}\exp\left(-\frac{(x-0.5)^2}{2}\right)\right),$$

$$\phi(x) = \sin(\pi x), \quad \psi(x) = 0, \quad w_0(t) = 0, \quad w_1(t) = 0, \quad U_T(x) = 0.$$

An inverse analysis was conducted to determine optimal values of mf and μ within the range (1–15), using the $CGLS$ algorithm with different tolerances and numbers of interior points N_x and N_t. The optimal parameters that yield accurate errors D and Z are as follows: Parameters: $\mu = 10$, $mf = 3$, $N_x = 11$, $N_t = 11$ Number of iterations: 315, with tol $= 10^{-10}$.

To assess the quality of our solutions we compared them with exact analytical solutions Figs. 5 and 6.

5.1 *Effect of Regularization and Preconditioning*

In solving the linear systems that arise in our methodology, we employed regularization techniques and TSkMP preconditioning methods to enhance numerical stability and convergence. To evaluate the impact of these strategies, we conducted a series of experiments. Table 1 presents the results of solving linear systems with and without regularization. The use of regularization techniques significantly improved the convergence and accuracy of our method, especially when dealing with ill-conditioned systems. After using the regularization and preconditioning and raring number of product N_p from (1–5) and (γ) from (0.1–1.0), the approximate errors D and Z are

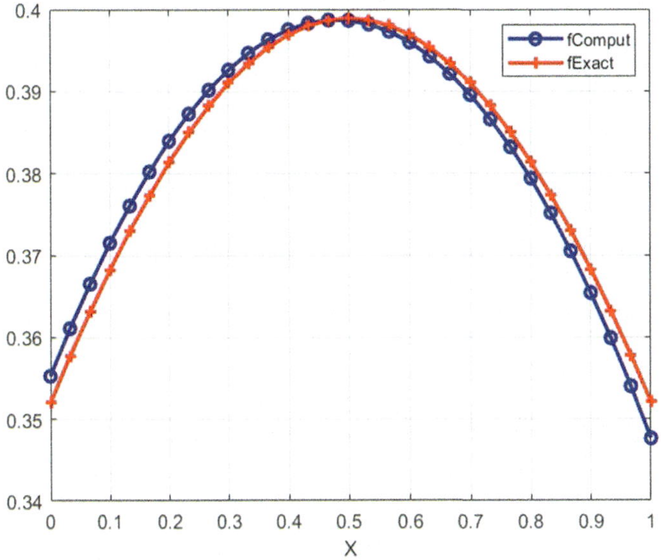

Fig. 4 Example 2: Exact and computed force without regularization or preconditioning

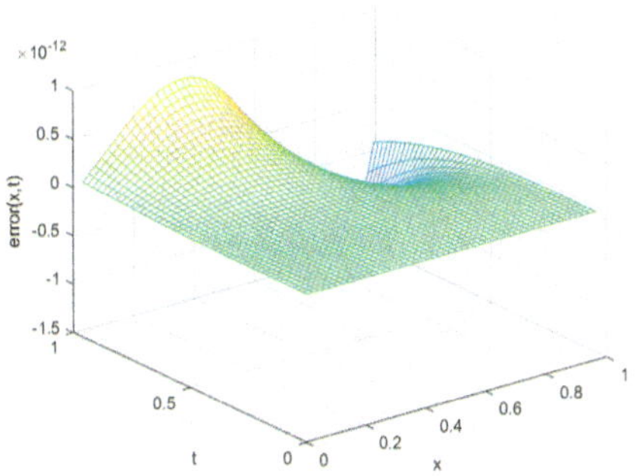

(a) Error of displacement u(x,t)

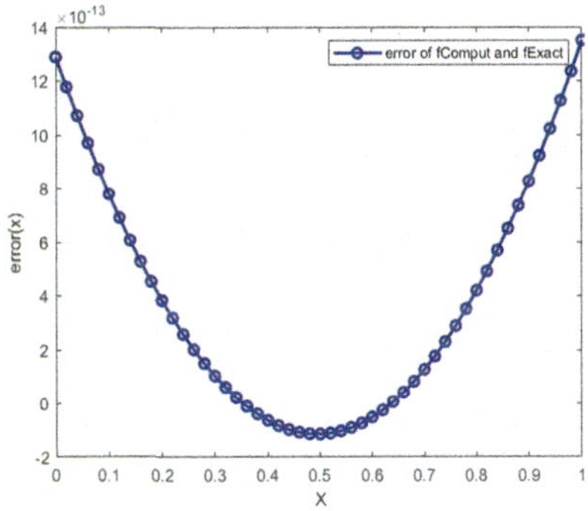

(b) Error of force f(x)

Fig. 5 Quality of solutions of Example 1

improved at ($\gamma = 0.6$) and ($N_p = 1$), for Example 1, and ($\gamma = 0.9$) and ($N_p = 1$), for Example 2.

Similarly, we explored the impact of preconditioning on the performance of our method. Figures 7 and 8 illustrates the convergence behavior of the iterative solver with and without preconditioning. The introduction of a suitable preconditioner led to accurate and faster convergence and a reduction in the number of iterations required to obtain solutions for Example 1, and Figs. 9 and 10 illustrates the convergence

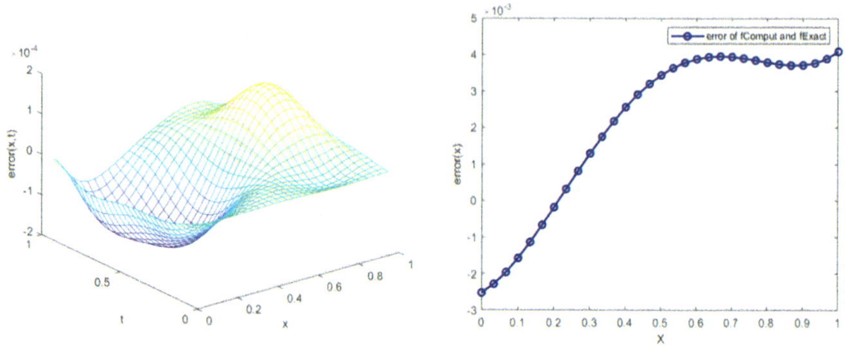

(a) Error of displacement u(x,t) (b) Error of force f(x)

Fig. 6 Quality of solutions of Example 2

Table 1 Effect of regularization and preconditioning

Example 1			
	Error (D)	Error (Z)	Iteration
Without regularization	1.299e–12	6.06e–11	81
With regularization and preconditioning	8.8485e–14	5.7521e–13	57
Example 2			
	Error (D)	Error (Z)	Iteration
Without regularization	2.2984e–4	4.5464e–3	315
With regularization and preconditioning	4.38e–4	6.366e–3	259

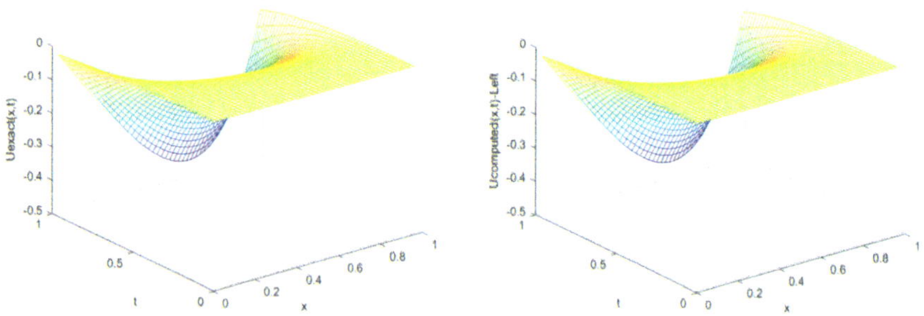

(a) Exact displacement u(x,t) (b) Computed displacement u(x,t)

Fig. 7 Example 1: Exact and computed displacement with regularization and preconditioning

behavior of the iterative solver with and without preconditioning. The introduction of a suitable preconditioner led to faster convergence and a reduction in the number of iterations required to obtain solutions for Example 2.

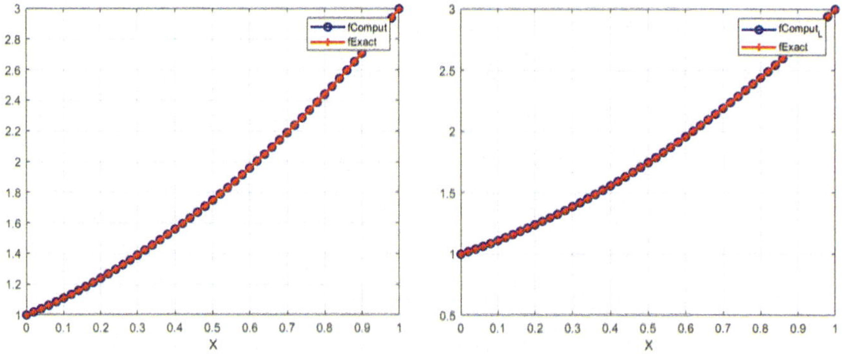

(a) No Regularization and no precondition (b) With Regularization and precondition

Fig. 8 Example 1: Exact and computed force and effect of regularization and preconditioning

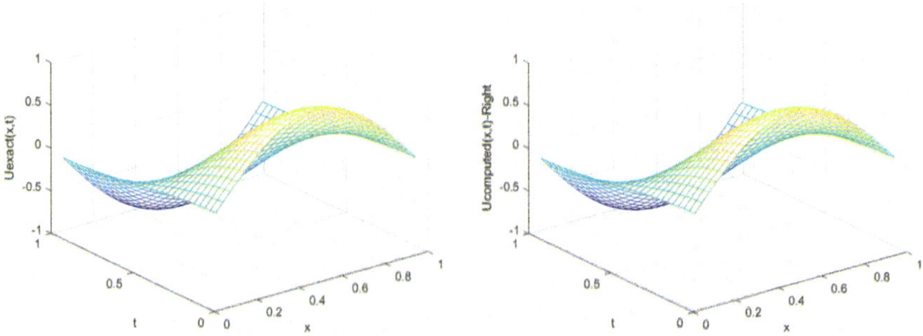

(a) Exact displacement (b) Computed displacement

Fig. 9 Example 2: Exact and computed displacement, with regularization and precondition

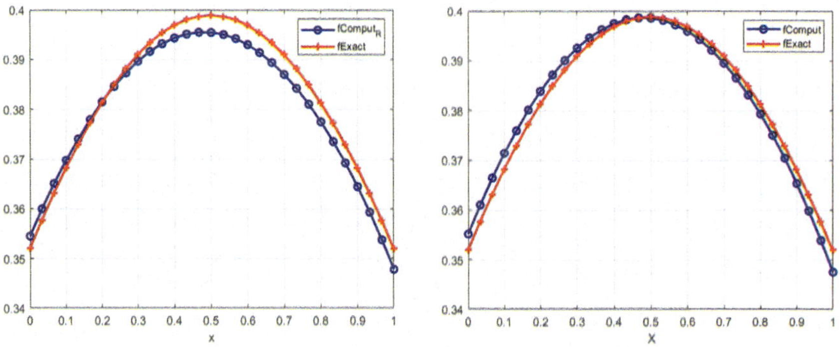

(a) No Regularization and no precondition (b) Regularization and precondition

Fig. 10 Example 2: Exact and computed force, effect of regularization and precondition

Table 2 Norm of error on the displacement for different noise levels

Order of noise	0.1	0.05	0.01	0.005	0.001
Norme of error	0.31084	0.14319	0.031905	0.015551	0.0030094

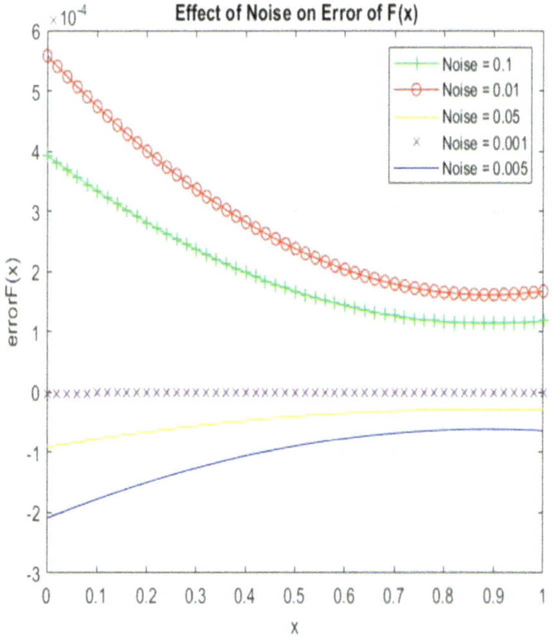

Fig. 11 Results of example (1) comparison of the effect of noise on its data

5.2 *Numerical Stability and Noise Tolerance*

To assess the numerical stability of our method and its tolerance to noise in the data, we conducted a robustness analysis. We intentionally introduced noise into the displacement data used for the inverse problem and observed the effect on the quality of the solutions. The results on Fig. 11 and Table 2 demonstrate that appropriate regularization, combined with multi-preconditioning techniques, can significantly enhance the stability and quality of solutions, even in the presence of noise within the data.

6 Conclusion

In this study, we investigated the performance of the polynomial expansion method for solving the problem of identifying of a space dependent force arising in various numerical applications.

Our results indicate that the proposed approach yields stable and accurate solutions, even in cases involving noisy data. The findings provide valuable insights into the application of regularization and preconditioning in improving the reliability of numerical solutions.

Future work could focus on expanding this approach to more complex systems and further exploring alternative preconditioning methods to improve convergence rates.

Acknowledgements This work was developed during the visit of Gulnar Wasim Sadiq, at the Laboratoire de Mathématiques Jean Leray, Nantes Université (France), CNRS UMR 6629. This scientific stay was partially funded by Le Service de Coopérationd'Action Culturelle (SCAC), Ambassade de France en Irak.

References

1. Aboud, F., Nachaoui, A., Nachaoui, M.: On the approximation of a Cauchy problem in a non-homogeneous medium. J. Phys.: Conf. Ser. **1743**(1), 012003 (2021)
2. Arsenashvili, A., Nachaoui, A., Tadumadze, T.: On approximate solution of an inverse problem for linear delay differential equations. Bull. Georgian Natl. Acad. Sci. (N.S.) **2**(2), 24–28 (2008)
3. Chakib, A., Johri, M., Nachaoui, A., Nachaoui, M.: On a numerical approximation of a highly nonlinear parabolic inverse problem in hydrology. An. Univ. Craiova Ser. Mat. Inform. **42**(1), 192–201 (2015)
4. Chakib, A., Nachaoui, A., Zeghal, A.: A shape optimization approach for an inverse heat source problem. Int. J. Nonlinear Sci. **12**(1), 78–84 (2012)
5. Chakib, A., Nachaoui, A., Nachaoui, M.: Approximation and numerical realization of an optimal design welding problem. Numer. Methods Partial Differ. Eq. **29**(5), 1563–1586 (2012)
6. Fang, W., Lu, M.: A fast collocation method for an inverse boundary value problem. Int. J. Numer. Methods Eng. **21**, 1563–1585 (2004)
7. Guliyev, H.F., Gasimov, Y.S., Tagiyev, H.T., Huseynova, T.M.: On the inverse problem of finding the right-hand side of wave equation with nonlocal condition. Vestn. Tomsk. Gos. Univ. Mat. Mekh **49**, 16–25 (2017)
8. Guliyev, H.F., Gasimov, Y.S., Zeynalli, S.M.: Optimal control method for solving the Cauchy-Neumann problem for the Poisson equation. Zh. Mat. Fiz. Anal. Geom. **12**(4), 412–421 (2014)
9. Isakov, V.: Inverse problems for partial differential equations. Appl. Math. Sci. **127** (2017)
10. Jin, B., Zheng, Y.: A meshless method for some inverse problems associated with the Helmholtz equation. Comput. Methods Appl. Mech. Eng. **195**(19–22), 2270–2288 (2006)
11. Jourhmane, M., Nachaoui, A.: An alternating method for an inverse Cauchy problem. Numer. Algorithms **21**(1), 247–260 (1999)
12. Nachaoui, M., Nachaoui, A., Tadumadze, T.: On the numerical approximation of some inverse problems governed by nonlinear delay differential equation. RAIRO Oper. Res. **56**(3), 1553–1569 (2022)

13. Nachaoui, A., Aboud, F.: Solving geometric inverse problems with a polynomial based meshless method. In: Springer Proceedings in Mathematics and Statistics, vol. 428, pp. 119–136. Springer, Cham (2023)
14. Nachaoui, A.: Cauchy's problem for the modified biharmonic equation: derivation and decoupled iterative methods. In: Springer Proceedings in Mathematics and Statistics, vol. 498. Springer, Cham (2025)
15. Reddy, G.M.M., Nanda, P., Vynnycky, M., Cuminato, J.A.: An adaptive boundary algorithm for the reconstruction of boundary and initial data using the method of fundamental solutions for the inverse Cauchy-Stefan problem. Comput. Appl. Math. **40**(3) (2021)
16. Aboud, F., Jameel, I.T., Hasan, A.F., Mostafa, B.K., Nachaoui, A.: Polynomial approximation of an inverse Cauchy problem for Helmholtz type equations. Adv. Math. Models Appl. **7**(3), 306–322 (2022)
17. Blum, J.: Numerical Simulation and Optimal Control in Plasma Physics with Applications to Tokamaks. Wiley, Chichester (1989)
18. Ellabib, A., Nachaoui, A., Ousaadane, A.: Convergence study and regularizing property of a modified robin-robin method for the Cauchy problem in linear elasticity. Inverse Probl. **38**(7), 075007 (2022)
19. Hernandez-Montero, E., Fraguela-Collar, A., Henry, J.: An optimal quasi solution for the Cauchy problem for Laplace equation in the framework of inverse ECG. Math. Model. Nat. Phenom. **14**(2) (2019)
20. Liu, J.C., Wei, T.: A quasi-reversibility regularization method for an inverse heat conduction problem without initial data. Appl. Math. Comput. **219**(23), 10866–10881 (2013)
21. Nachaoui, A., Nachaoui, M., Tadumadze, T.: Meshless methods to noninvasively calculate neurocortical potentials from potentials measured at the scalp surface. In: Springer Proceedings in Mathematics & Statistics, vol. 428, pp. 99–117. Springer, Cham (2023)
22. Maxwell, D., Truffer, M., Avdonin, S., Stueferv, M.: An iterative scheme for determining glacier velocities and stresses. J. Glaciol. **54**(188), 888–898 (2008)
23. Dvalishvili, P., Nachaoui, A., Nachaoui, M., Tadumadze, T.: On the well-Posedness of the Cauchy problem for a class of differential equations with distributed delay and the continuous initial condition. Proc. Inst. Math. Mech. Natl. Acad. Sci. Azerb. **34**(1), 146–160 (2017)
24. Hadamard, J.: Lectures on Cauchy's Problem in Linear Partial Differential Equations. Yale University Press, New Haven (1923)
25. Lavrentiev, M.M.: Some Improperly Posed Problems of Mathematical Physics. SIAM, PhiladelphiaSpringer Science and Business Media (2013)
26. Cannon, J.R., Dunninger, D.R.: Determination of an unknown forcing function in a hyperbolic equation from overspecified data. Ann. Mat. Pura Appl. **1**, 49–62 (1970)
27. Yamamoto, M.: Stability, reconstruction formula and regularization for an inverse source hyperbolic problem by a control method. Inverse Probl. **11**, 481–496 (1995)
28. Engl, H.W., Scherzer, O., Yamamoto, M.: Uniqueness and stable determination of forcing terms in linear partial differential equations with overspecified boundary data. Inverse Probl. **10**, 1253–1276 (1994)
29. Pilant, M.S., Rundell, W.: Undetermined coefficient problems for quasilinear parabolic equations. In: Inverse Problems in Partial Differential Equations, SIAM, Philadelphia, pp. 165–185 (1990)
30. Prilepko, A.I., Orlovsky, D.G., Vasin, I.A.: Methods for Solving Inverse Problems in Mathematical Physics. Marcel Dekker Inc, New York (2000)
31. Hasanov, A.: Simultaneous determination of the source terms in a linear hyperbolic problem from the final overdetermination: weak solution approach. IMA J. Appl. Math. **74**, 1–19 (2009)
32. Slodicka, M.: Some uniqueness theorems for inverse spacewise dependent source problems in nonlinear PDES. Inverse Probl. Sci. Eng. **22**, 2–9 (2014)
33. Nachaoui, A., Sadiq, G.W.: Investigation of an inverse problem in heat propagation. To appear (2025)
34. Nachaoui, A.: An improved implementation of an iterative method in boundary identification problems. Numer. Algorithms **33**(1–4), 381–398 (2003)

35. Lavrentiev, M.M.: Some Improperly Posed Problems of Mathematical Physics. Springer Science & Business Media (2013)
36. Huang, C.H., Chen, W.C.: A three-dimensional inverse forced convection problem in estimating surface heat flux by conjugate gradient method. Int. J. Heat Mass Transf. **43**(17), 3171–3181 (2000)
37. Boulkhemair, A., Nachaoui, A., Chakib, A.: A shape optimization approach for a class of free boundary problems of Bernoulli type. Appl. Math. **58**(2), 205–221 (2013)
38. Chakib, A., Hadri, A., Laghrib, A.: On a multiscale analysis of an inverse problem of nonlinear transfer law identification in periodic microstructure. Nonlinear Anal. Real World Appl. **51**, 102985 (2020)
39. Wang, F., Gu, Y., Qu, W., Zhang, C.: Localized boundary knot method and its application to large-scale acoustic problems. Comput. Methods Appl. Mech. Eng. **361**, 112729 (2020)
40. Nachaoui, A., Laghrib, A., Hakim, M.: Mathematical control and numerical applications. In: Springer Proceedings in Mathematics and Statistics, vol. 372. Springer, Cham (2021)
41. Nachaoui, A., Salih, H.W.: An analytical solution for the nonlinear inverse Cauchy problem. Adv. Math. Models Appl. **6**(3), 191–206 (2021)
42. Afraites, L., Masnaoui, C., Nachaoui, M.: Shape optimization method for an inverse geometric source problem and stability at critical shape. Discrete Contin. Dyn. Syst. Ser. S **15**(1), 1–21 (2022)
43. Nachaoui, A., Rashid, S.M.: A mesh free wavelet method to solve the Cauchy problem for the Helmholtz equation. In: Springer Proceedings in Mathematics and Statistics, vol. 428, pp. 77–98. Springer, Cham (2023)
44. Hadri, A., Laghrib, A., Nachaoui, A., Nachaoui, M.: On the resolution of the non-smooth inverse Cauchy problem by the primal-dual method. Numer. Algorithms (2024). URL https://doi.org/10.1007/s11075-024-01950-5
45. Nachaoui, A., Rasheed, S.M., Sadiq, G.W.: A semi-analytical solution for inverse Cauchy problems. In: Springer Proceedings in Mathematics & Statistics, vol. 498. Springer, Cham (2025)
46. Vasil'ev, V.V.P.V.I., Kardashevsky, A.M., Prokopev, G.A.: Iterative solution of the inverse Cauchy problem for an elliptic equation by the conjugate gradient method. AIP Conf. Proceed. **1895**(1), 110009 (2017)
47. Berdawood, K., Nachaoui, A., Nachaoui, M.: An accelerated alternating iterative algorithm for data completion problems connected with Helmholtz equation. Stat. Optim. Inf. Comput. **11**(1), 2–21 (2023). https://doi.org/10.1002/num.22793
48. Nachaoui, A., Aboud, F., Nachaoui, M.: Acceleration of the KMF algorithm convergence to solve the Cauchy problem for Poisson's equation. In: Springer Proceedings in Mathematics and Statistics, pp. 43–57. Springer, Cham (2021)
49. Lattès, R., Lions, J.: The method of quasi-reversibility: applications to partial differential equations. In: Modern Analytic and Computational Methods in Science and Mathematics. American Elsevier Publishing Co., Inc, New York (1969)
50. Cheng, X., Gong, R., Han, W.: A coupled complex boundary method for the Cauchy problem. Inverse Probl. Sci. Eng. **24**(9), 1510–1527 (2016)
51. Nachaoui, A., Gontier, L.: Analytical solutions for inverse Cauchy problems: a comparative study. In: Springer Proceedings in Mathematics and Statistics, vol. 498. Springer, Cham (2025)
52. Shen, S.N.A., Atluri, S.N.: The Meshless Local Petrov-Galerkin (MLPG) Method (2002)
53. Belytschko, T., Krongauz, Y., Organ, D., Fleming, M., Krysl, P.: Meshless methods: an overview and recent developments. Comput. Methods Appl. Mech. Eng. **139**, 3–47 (1996)
54. Fairweather, G., Karageorghis, A.: The method of fundamental solutions for elliptic boundary value problems. Adv. Comput. Math. **9**, 69–95 (1998)
55. Hua, Q., Gu, Y., Qu, W., Chen, W., Zhang, C.: A meshless generalized finite difference method for inverse Cauchy problems associated with three-dimensional inhomogeneous helmholtz-type equations. Eng. Anal. Bound. Elem. **82**, 162–171 (2017)
56. Rasheed, S.M., Nachaoui, A., Hama, M.F., Jabbar, A.K.: Regularized and preconditioned conjugate gradient like-methods methods for polynomial approximation of an inverse cauchy problem. Adv. Math. Models Appl. **6**(2), 89–105 (2021)

57. Wei, T., Hon, Y.C., Ling, L.: Method of fundamental solutions with regularization techniques for Cauchy problems of elliptic operators. Eng. Anal. Bound. Elem. **31**(4), 373–385 (2007)
58. Yue, X., Wang, F., Hua, Q., Qiu, X.: A novel space-time meshless method for nonhomogeneous convection-diffusion equations with variable coefficients. Appl. Math. Lett. **92**, 144–150 (2019)
59. Liu, C.S., Atluri, S.N.: A highly accurate technique for interpolations using very high-order polynomials, and its applications to some ill-posed linear problems. Comput. Model. Eng. Sci. **43**(1), 253–276 (2009)
60. Rashid, S.M., Nachaoui, A.: A Haar wavelets-based direct reconstruction method for the Cauchy problem of the Poisson equation. Discrete Contin. Dyn. Syst. Ser. S **18**(1), 148–168 (2025)
61. Boumzough, K., Azzouzi, A., Bouidi, A.: The incomplete lu preconditoner using both csr and csc formats. Adv. Math. Models Appl. **7**(2), 156–167 (2022)
62. Ye, Q.: Preconditioning for accurate solutions of ill-conditioned linear systems. Numer. Linear Algebra Appl. **27**(4), e2315 (2020)

Cauchy's Problem for the Modified Biharmonic Equation: Derivation and Decoupled Iterative Methods

Abdeljalil Nachaoui

Abstract This paper addresses the Cauchy problem for the Modified Biharmonic Equation (MBE). We start by presenting its derivation from the Stokes equations under specific assumptions and boundary conditions. A factorization technique is introduced to manage this fourth-order Cauchy problem, leading to the development of an iterative alternating method for solving the resulting second-order Cauchy problems. The finite element method is used for numerical discretization. The main findings are summarized and analyzed in the context of numerical experiments, with a discussion on their significance. Finally, we outline potential extensions and the limitations of this study.

Keywords Modified biharmonic equation · Inverse cauchy problem · Iterative regularization

1 Introduction

Numerous linear elliptic equations involving compositions of second-order elliptic differential operators, such as the biharmonic operator ∇^4, frequently emerge in various models and describe a wide array of phenomena in applied sciences [5, 18, 27, 29].

The development of the theory of inverse and ill-posed problems for the biharmonic equation has significantly advanced in recent years, driven by the capabilities of modern computational power [27]. Among the most studied inverse problems is the Cauchy problem [2, 12, 14, 19, 20, 24, 37, 45, 49], which involves reconstructing unknown data on an inaccessible part of the boundary from known Cauchy data on another portion. The investigation of the Cauchy problem is motivated by its extensive practical applications in fields such as fluid dynamics and engineering.

A. Nachaoui (✉)
Laboratoire de Mathématiques Jean Leray, Nantes Université, Nantes, France
e-mail: Abdeljalil.Nachaoui@univ-nantes.fr

A. Laghrib et al. (eds.), *Computational Methods for Inverse Problems and Applications*, Springer Proceedings in Mathematics & Statistics 498,
https://doi.org/10.1007/978-3-031-89498-5_10

A notable application lies in medical imaging, particularly in estimating neurocortical potentials from scalp measurements, where solving the forward problem links the sources to observed data, and solving the inverse problem reconstructs these sources [38].

Another significant area of application is fluid flow in bounded domains. Under certain conditions, the modified biharmonic equation serves as a model for the flow of a viscous fluid. Solving the Cauchy problem in this context yields valuable insights into the velocity and pressure fields, which hold critical importance in various engineering applications.

Various methods have been developed to address inverse problems in applied science [3, 4, 9, 11, 22, 28, 31, 35, 39, 40, 42–44]. One of the most widely employed techniques for solving Cauchy problems is the relaxed alternating iterative method. This approach simplifies the resolution process by alternately solving two well-posed sub-problems [25].

This iterative procedure significantly reduces the number of iterations required to achieve convergence in the inverse problems being studied. Recent studies have substantiated these convergence acceleration results, notably for the Poisson equation [34] and the Helmholtz equation [6]. The method has also been adapted to nonlinear elliptic problems, as shown in [16, 17]. The versatility of these techniques allows for applications across a broad spectrum of inverse boundary value problems, including nonlinear elliptic problems [16, 17], structural mechanics [12–15], electromagnetism [7], fluid mechanics, and problems involving non-homogeneous media [2]. Additionally, it has been applied to the convection-diffusion equation [33].

Aside from the relaxed alternating iterative method, numerous other strategies have been developed for tackling inverse problems. Comprehensive discussions and further developments on these methods can be found in the literature, such as [1, 21, 26, 30, 32, 41, 46–48, 50, 51] and the references therein. These contributions continue to enrich the field by offering alternative approaches and enhancing the robustness and applicability of inverse problem-solving techniques across various disciplines.

We focus on the Cauchy problem governed by a modified biharmonic equation. The problem assumes that Cauchy data is available on a portion of the boundary and the solution needs to be reconstructed on an inaccessible section of the boundary. The objective is to develop an effective numerical method to solve this problem.

A powerful approach for the numerical solution of the biharmonic equation is the mixed finite element method [10]. This method is particularly well-suited for handling biharmonic equations with Navier boundary conditions. The advantage of employing the mixed finite element method [8, 36] lies in its ability to reformulate a fourth-order problem into two decoupled second-order equations. This decomposition enables the application of well-established finite element Poisson solvers, thereby simplifying the computational process and improving efficiency.

In this study, we aim to leverage this principle to develop an iterative method for the numerical resolution of the modified biharmonic Cauchy problem. By adopting this approach, we ensure that the numerical scheme is robust and capable of handling the intricacies of the boundary conditions while maintaining accuracy and convergence.

2 Derivation of the Modified Biharmonic Equation

The time-dependent, linearized, viscous, incompressible flow (commonly known as unsteady Stokes flow or the linearized Navier-Stokes equations), which describes the motion of a viscous fluid within a bounded domain $\Omega \subset \mathbb{R}^n$ with a smooth boundary $\partial\Omega$ for $t \in (0, T)$ under the influence of external forces, can be represented in vector form as:

$$\nabla \cdot \boldsymbol{u} = 0 \qquad \text{in } \Omega \times (0, T) \tag{1}$$

$$\frac{\partial \boldsymbol{u}(x,t)}{\partial t} = \boldsymbol{f}(x,t) + \nabla p + \nu \nabla^2 \boldsymbol{u}(x,t) \qquad \text{in } \Omega \times (0, T), \tag{2}$$

where $\boldsymbol{u}$ is the velocity vector, p is the pressure, $\boldsymbol{f}$ represents the external force per unit volume, ν is the kinematic viscosity, and ∇ denotes the gradient operator. Depending on the specific problem, the forcing term $\boldsymbol{f}$ may be zero or a known function. The full incompressible Navier-Stokes equations can also be represented in the form of (2) by setting $\boldsymbol{f} = -\boldsymbol{u} \cdot \nabla \boldsymbol{u}$.

To reduce the system of unknowns from three to one, we introduce the scalar stream function ψ, which relates to the velocity field through the relation $\boldsymbol{u} = \nabla \times (k\psi)$, where k is the unit vector perpendicular to the plane in $\mathbb{R}^2$. Consequently, the components of the velocity field can be expressed as:

$$u = \frac{\partial \psi}{\partial y}, \qquad v = -\frac{\partial \psi}{\partial x}, \tag{3}$$

where u and v represent the x and y components of the velocity vector, respectively. By taking the curl of (2) to eliminate the pressure term, we obtain the equation:

$$\left(\frac{\partial}{\partial t} - \frac{1}{Re}\nabla^2\right) \nabla^2 \psi(x,t) = -\nabla \times \boldsymbol{f}(x,t) \cdot k, \tag{4}$$

where Re denotes the Reynolds number.

To solve this equation numerically, we first discretize (4) in time. The biharmonic term is treated implicitly to avoid stringent time-step restrictions. Using the first-order backward Euler method for temporal discretization, we derive the modified biharmonic equation as follows: To discretize the given equation in time using the first-order backward Euler method, we can approximate the time derivative as follows:

$$\frac{\partial}{\partial t}\nabla^2 \psi(x,t) \approx \frac{1}{\delta t}\left(\nabla^2 \psi(x, t^n) - \nabla^2 \psi(x, t^{n-1})\right)$$

where δt represents the time step size, and t^n and t^{n-1} denote the current and previous time levels, respectively.

Substituting this approximation into the original equation yields:

$$\frac{1}{\delta t}\left(\nabla^2\psi(x,t^n)-\nabla^2\psi(x,t^{n-1})\right)-\frac{1}{Re}\nabla^4\psi(x,t^n)=-\nabla\times \boldsymbol{f}(x,t^{n-1})\cdot k$$

Rearranging the terms results in the following expression for $\nabla^2\psi(x,t^n)$:

$$(\nabla^2-\lambda^2)\nabla^2\psi(x,t^n)=b(x,t^n),\quad (x,y)\in\Omega, \tag{5}$$

where $\lambda^2=\frac{\delta t}{Re}$ and $b(x,t^{n-1})=\nabla^2\psi(x,t^{n-1})+\delta t\left(\nabla\times \boldsymbol{f}(x,t^{n-1})\cdot k\right)$.

This modified biharmonic equation connects the stream function ψ with the vorticity and external force $\boldsymbol{f}$. It incorporates the fourth-order bi-Laplacian operator ∇^4, which is highly dissipative and can pose numerical challenges.

This equation provides a method to compute $\psi(x,t^n)$ at each discrete time step, based on the previous value $\psi(x,t^{n-1})$ and the forcing term $\nabla\times\boldsymbol{f}(x,t^{n-1})\cdot k$. Once $\psi(x,t^n)$ has been calculated at all spatial grid points and time steps, it can be used to derive the velocity field $\boldsymbol{u}(x,t)$ through the relations in equations (3). At each time step, we can express the solution as $\psi(x,t^n)=\psi^p+u$, where ψ^p is any solution to Eq. (5), and u satisfies the following equation:

$$(\nabla^2-\lambda^2)\nabla^2u(x)=0,\quad (x)\in\Omega. \tag{6}$$

Note that in the context of integral equations, ψ^p is simply the volume integral of the nonhomogeneous term multiplied by the green function, allowing for direct evaluation without requiring a solution process [23].

Since Eq. (6) is a fourth-order elliptic partial differential equation, it necessitates two boundary conditions for the solution to be unique: one boundary condition on u (either Dirichlet or Neumann) and another condition on Δu (also Dirichlet or Neumann).

$$u \text{ or } \partial_\nu u=f \text{ on } \partial\Omega \qquad \text{and} \qquad \Delta u \text{ or } \partial_\nu\Delta u=g \text{ on } \partial\Omega,$$

where $\partial_\nu=\frac{\partial}{\partial\nu}$ denotes the outward normal derivative.

In many scenarios, boundary conditions may only be specified on part of the boundary, requiring reconstruction of data for inaccessible sections based on Cauchy data available on accessible parts.

Consider the following biharmonic equation, along with the associated eigenvalue problem:

$$\Delta^2\phi=f \text{ in } \Omega,\qquad \phi=\Delta\phi=0 \text{ on } \partial\Omega; \tag{7}$$

where $f\in L^2(\Omega)$. The weak formulation of the biharmonic problem (7) seeks $\phi\in H_0^1(\Omega)\cap H^2(\Omega)$ such that

$$(\Delta\phi,\Delta\psi)=(f,\psi)\quad \forall\psi\in H_0^1(\Omega)\cap H^2(\Omega), \tag{8}$$

These formulations require the application of H^2-conforming methods, which can be quite complex. In practice, this approach may not be desirable. Instead of using such methods, a widely adopted alternative is to implement a mixed formulation.

Typically, we can set $\bar{\rho} = -\Delta\bar{\phi}$, allowing us to reformulate Eq. (7) into the following mixed formulation: find $(\bar{\rho}, \bar{\phi}) \in H_0^1(\Omega) \times H_0^1(\Omega)$ such that

$$(\nabla\bar{\phi}, \nabla\psi) = (\bar{\rho}, \psi), \quad \forall\psi \in H_0^1(\Omega), \tag{9a}$$

$$(\nabla\bar{\rho}, \nabla\nu) = (f, \nu), \quad \forall\nu \in H_0^1(\Omega). \tag{9b}$$

We observe that the system in (9) comprises two completely decoupled Poisson equations, making it significantly simpler to solve than the original biharmonic equation (7). This mixed formulation provides greater flexibility and can facilitate more efficient numerical solutions, especially in cases where standard conforming methods may encounter difficulties.

3 Cauchy Problem Governed by the Modified Biharmonic Equation

The Cauchy problem is a mathematical challenge aimed at finding a solution to a partial differential equation that adheres to specified boundary conditions. For the modified biharmonic equation, the formulation of the Cauchy problem is as follows: Given a bounded domain Ω with a piecewise smooth boundary $\partial\Omega$, let Γ_1 and Γ_2 denote two disjoint segments of the boundary such that $\partial\Omega = \Gamma_1 \cup \Gamma_2$. We are provided with functions $f_1(\boldsymbol{x})$, $f_1'(\boldsymbol{x})$, $h_1(\boldsymbol{x})$, and $h_1'(\boldsymbol{x})$ defined on Γ_1. The boundary conditions are specified solely on Γ_1 as follows:

$$u = f_1 \quad \text{and} \quad \partial_\nu u = f_1' \quad \text{for } x \in \Gamma_1,$$

$$\Delta u = h_1 \quad \text{and} \quad \partial_\nu \Delta u = h_1' \quad \text{for } x \in \Gamma_1.$$

The boundary conditions on Γ_2 are unknown, necessitating reconstruction from the given data on Γ_1. This reconstruction is essential to ensure the existence of a solution to the Cauchy problem. The uniqueness of the solution, when it exists, can be established using a continuation argument (Fig. 1).

To illustrate the problem, let Ω be a rectangle in $\mathbb{R}^2$ defined by the sides $\Sigma_1 = [x_a, x_b] \times \{y_a\}$, $\Gamma_m = \{x_b\} \times [y_a, y_c]$, $\Sigma_3 = [x_a, x_b] \times \{y_c\}$, and $\Gamma_u = \{x_a\} \times [y_a, y_c]$, where $x_a, y_a, x_b, y_c \in \mathbb{R}$.

We seek to solve the following Cauchy problem governed by the modified biharmonic equation:

Fig. 1 Simulation domain

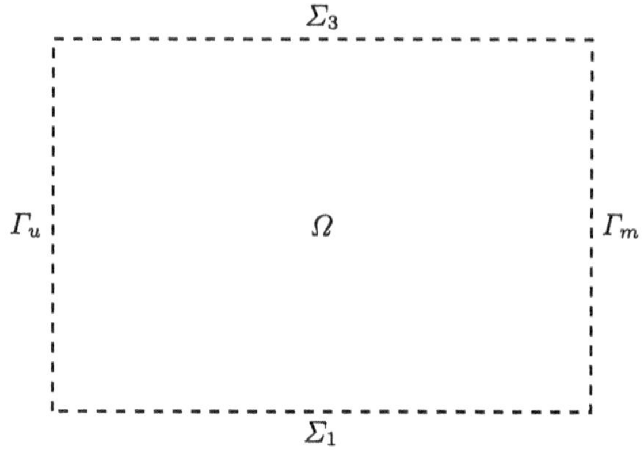

Find u such that:

$$\begin{cases} L_1(L_2(u)) = 0 \text{ in } \Omega \\ u = h_1, L_2(u) = \omega_1 \text{ on } \Sigma_1 \text{ and } \partial_\nu u = g_3, \partial_\nu L_2(u) = \omega_3' \text{on } \Sigma_3 \\ u = h_2, L_2(u) = \omega_2 \text{ on } \Gamma_m \text{ and } \partial_\nu u = g_2, \partial_\nu L_2(u) = \omega_2' \text{on } \Gamma_m \end{cases} \tag{10}$$

where $L_1(u) = (\Delta - \lambda)(u)$ and $L_2(u) = \Delta(u)$, with λ being a positive constant.

Following the mixed formulation approach (9a)-(9b), we can express the problem (10) in terms of the operator L_2 as follows:

$$\begin{cases} L_1(v) = f \text{ in } \Omega & \text{and } L_2(u) = v \text{ in } \Omega \\ v = \omega_1 \text{ on } \Sigma_1; \partial_\nu v = \omega_3' \text{ on } \Sigma_3 & \text{and } u = h_1 \text{ on } \Sigma_1; \partial_\nu u = g_3 \text{ on } \Sigma_3 \\ v = \omega_2, \partial_\nu v = \omega_2' \text{ on } \Gamma_m & \text{and } u = h_2, \partial_\nu u = g_2 \text{ on } \Gamma_m \end{cases} \tag{11}$$

We can apply this factorization to obtain:

$$\begin{cases} & L_1(v) = f, \text{ in } \Omega & \\ v = \omega_1, \text{ on } \Sigma_1 & \text{and} & \partial_\nu v = \omega_3' \text{ on } \Sigma_3 \\ v = \omega_2 \text{ on } \Gamma_m & \text{and} & \partial_\nu v = \omega_2' \text{ on } \Gamma_m \end{cases} \tag{12}$$

and

$$\begin{cases} & L_2(u) = v \text{ in } \Omega & \\ u = h_1 \text{ on } \Sigma_1 & \text{and} & \partial_\nu u = g_3 \text{on } \Sigma_3 \\ u = h_2 \text{ on } \Gamma_m & \text{and} & \partial_\nu u = g_2 \text{on } \Gamma_m \end{cases} \tag{13}$$

The fourth-order Cauchy problem governed by the modified biharmonic equation (1) is decomposed into two second-order Cauchy subproblems governed by a Helmholtz equation and a Poisson equation. These second-order Cauchy problems are only coupled at the level of the first equation of the second problem, where the solution from the first problem serves as the source term for the second. Therefore, we can decouple the resolution of these problems by solving them sequentially.

To begin, we address the Cauchy problem for the modified Helmholtz equation (12) using standard numerical methods. Once the solution v to (12) is obtained, we proceed to determine u by solving the Cauchy problem for the Poisson equation (13). The solution from (12) serves as input for this step, and numerical techniques are applied once more to achieve the solution.

This decoupling facilitates the application of methods for solving Cauchy problems associated with second-order elliptic equations. In this regard, we utilize the alternating iterative method developed by the author [24], which has been successfully applied in various domains [2, 6, 7, 12, 13, 16, 17, 33, 34, 41]. One advantage of this method is that both problems are treated equivalently since the method does not alter the governing equation; only the boundary condition on the inaccessible boundary Γ_u is updated iteratively, following the same approach for both problems.

4 An Iterative Alternating Method

The two problems (12) and (13) can be seen as Cauchy problems, with the first governed by a Helmholtz equation and the second associated with the Poisson equation. We can express these problems in a unified form:

Find φ such that

$$\begin{cases} \qquad\qquad L(\varphi) = F \text{ in } \Omega \\ \varphi = \phi_1 \text{ on } \Sigma_1 \quad \text{and} \quad \partial_\nu \varphi = \psi_3 \text{ on } \Sigma_3 \\ \varphi = \phi_m \text{ on } \Gamma_m \quad \text{and} \quad \partial_\nu \varphi = \psi_m \text{ on } \Gamma_m \end{cases} \tag{14}$$

where the operator L may be either L_1 or L_2, and the boundary conditions correspond to those of problems (12) or (13), respectively.

We define the following two well-posed problems:

Find φ such that:

$$\begin{cases} \qquad\qquad L(\varphi) = F \text{ in } \Omega \\ \varphi = h \text{ on } \Gamma_u \quad \text{and} \quad \varphi = \phi_1 \text{ on } \Sigma_1 \\ \partial_\nu \varphi = \psi_m \text{ on } \Gamma_m \quad \text{and} \quad \partial_\nu \varphi = \psi_3 \text{ on } \Sigma_3 \end{cases} \tag{15}$$

and

Find φ such that:

$$\begin{cases} \qquad\qquad L(\varphi) = F \text{ in } \Omega \\ \partial_\nu \varphi = g \text{ on } \Gamma_u \quad \text{and} \quad \varphi = \phi_1 \text{ on } \Sigma_1 \\ \varphi = \phi_m \text{ on } \Gamma_m \quad \text{and} \quad \partial_\nu \varphi = \psi_3 \text{ on } \Sigma_3 \end{cases} \tag{16}$$

The alternating iterative algorithm to solve the Cauchy problem (14) is as follows:

Algorithm 1: JN Iterative Alternating Method [24]

1: $n \leftarrow 0$ take $h \leftarrow h^0$ arbitrary, $\theta \in (0, 2]$ and $\varepsilon > 0$
2: Find $\varphi^{(2n)}$, the solution of (15)
3: $g \leftarrow \partial_\nu \varphi^{(2n)}_{|\Gamma_u}$
4: Find $\varphi^{(2n+1)}$, the solution of (16)
5: $e^{(n)} \leftarrow \|\varphi^{(2n+1)} - \varphi^{(2n-1)}\|_{L^2(\Gamma_u)}$
6: If $e^{(n)} > \varepsilon$ then $h \leftarrow \theta\varphi^{(2n+1)} + (1-\theta)\varphi^{(2n)}$, $n \leftarrow n+1$ and go to step 2.

By applying this algorithm successively to the two Cauchy problems (12) and (13), we derive the following algorithm for solving the Cauchy problem governed by the modified biharmonic equation (11).

Algorithm 2: Nachaoui's Successive Iterative Alternating Method

1: $L \leftarrow L_1$, $F \leftarrow f$, $\phi_1 \leftarrow \omega_1$, $\phi_m \leftarrow \omega_2$, $\psi_m \leftarrow \omega_2'$, $\psi_3 \leftarrow \omega_3'$,
2: $n \leftarrow 0$, take $h \leftarrow h^0$ arbitrary, $\theta \in (0, 2]$ and $\varepsilon > 0$
3: Find $\varphi^{(2n)}$, the solution of (15)
4: $g \leftarrow \partial_\nu \varphi^{(2n)}_{|\Gamma_u}$
5: Find $\varphi^{(2n+1)}$, the solution of (16)
6: $e^{(n)} \leftarrow \|\varphi^{(2n+1)} - \varphi^{(2n-1)}\|_{L^2(\Gamma_u)}$
7: If $e^{(n)} > \varepsilon$ then $h \leftarrow \theta\varphi^{(2n+1)} + (1-\theta)\varphi^{(2n)}$, $n \leftarrow n+1$ and go to step 3.
8: $v \leftarrow \varphi^{(2n+1)}$
9: $L \leftarrow L_2$, $F \leftarrow v$, $\phi_1 \leftarrow h_1$, $\phi_m \leftarrow h_2$, $\psi_m \leftarrow g_2$, $\psi_3 \leftarrow g_3$,
10: $n \leftarrow 0$, take $h \leftarrow h^0$
11: Find $\varphi^{(2n)}$, the solution of (15)
12: $g \leftarrow \partial_\nu \varphi^{(2n)}_{|\Gamma_u}$
13: Find $\varphi^{(2n+1)}$, the solution of (16)
14: $e^{(n)} \leftarrow \|\varphi^{(2n+1)} - \varphi^{(2n-1)}\|_{L^2(\Gamma_u)}$
15: If $e^{(n)} > \varepsilon$ then $h \leftarrow \theta\varphi^{(2n+1)} + (1-\theta)\varphi^{(2n)}$, $n \leftarrow n+1$ and go to step 11
16: $u \leftarrow \varphi^{(2n+1)}$

The decoupling discussed in the previous section enables us, through Algorithm 1, to construct a sequence of solutions for well-posed problems, whose limit corresponds to the solution of the Cauchy problem (11). In the subsequent section, we present numerical results demonstrating this convergence.

5 Numerical Results and Discussion

In the numerical experiments, we employed the finite element method $\mathbb{P}_1$ to approximate the two problems (15) and (16), using a grid of 25 points in the x direction and 100 points in the y direction. The initial guess was set to $h^{(0)} = 0$. We chose the parameter $\lambda = 1$.

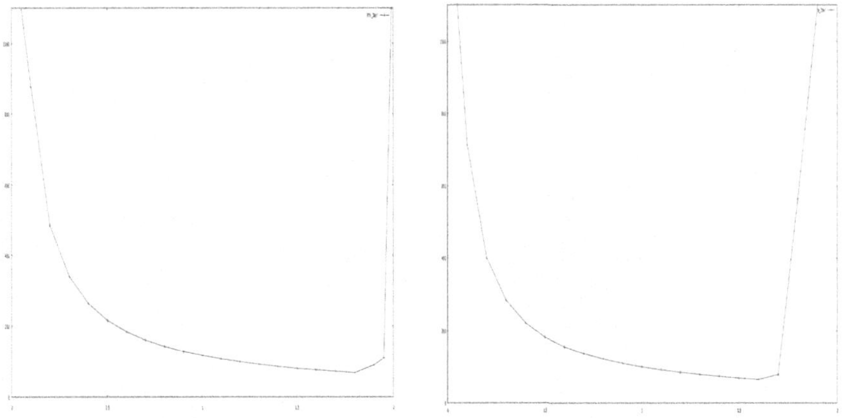

(a) (b)

Fig. 2 Example 1, iterations as a fuction of θ: (2a) for L_1, (2b) for L_2

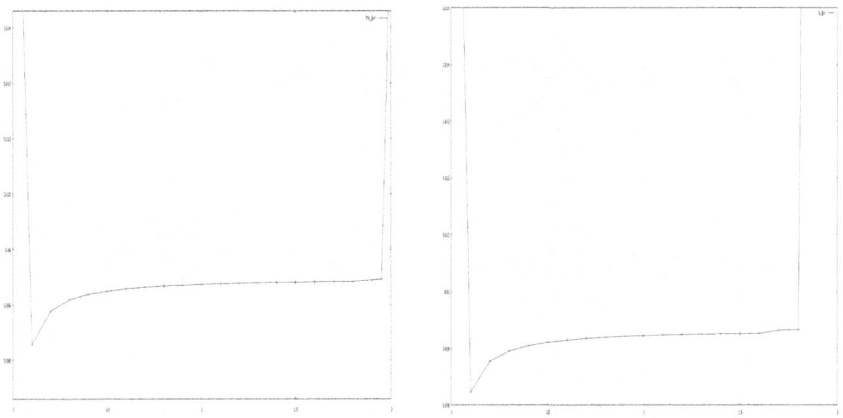

(a) (b)

Fig. 3 Example 1, Errors as a fuction of θ: (2a) for L_1, (2b) for L_2

Example 1 The first example considers data derived from the analytical solution given by $u(x, y) = y^2 + xy + x + y + 1$. The initial guess remains $h^{(0)} = 0$.

From Fig. 2, we observe that the number of iterations increases significantly as the relaxation parameter approaches the boundaries of the interval $(0, 2)$, leading to slower convergence in this region. Additionally, each of the two Cauchy problems exhibits an optimal parameter that minimizes the number of iterations required.

The optimal parameters differ for the two problems, which is logical since the first problem is governed by the Helmholtz equation (Fig. 2a), while the second is governed by the Poisson equation (Fig. 2b).

Figure 3 illustrates that the error norm follows a similar trend to that of the iterations. Notably, as convergence slows, the calculated solution increasingly diverges

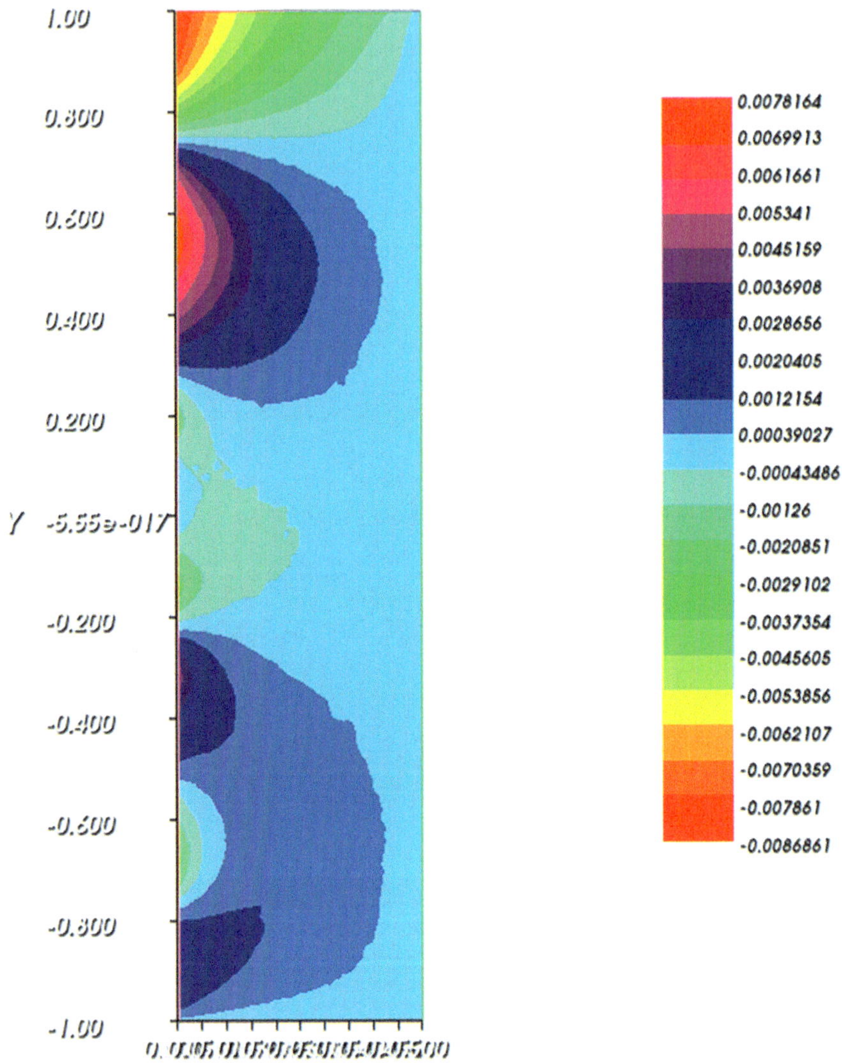

Fig. 4 Error for Example 1

from the exact solution. This stall is abrupt after the parameter that provides the fastest convergence.

This proximity is confirmed in Fig. 4 shows that the error remains globally below 4×10^{-3}, with a maximum error near the inaccessible boundary, yet still under 8.6×10^{-3}. Figure 5 illustrates that even at Γ_u, where the error peaks, the graphs of the exact and computed solutions are indistinguishable.

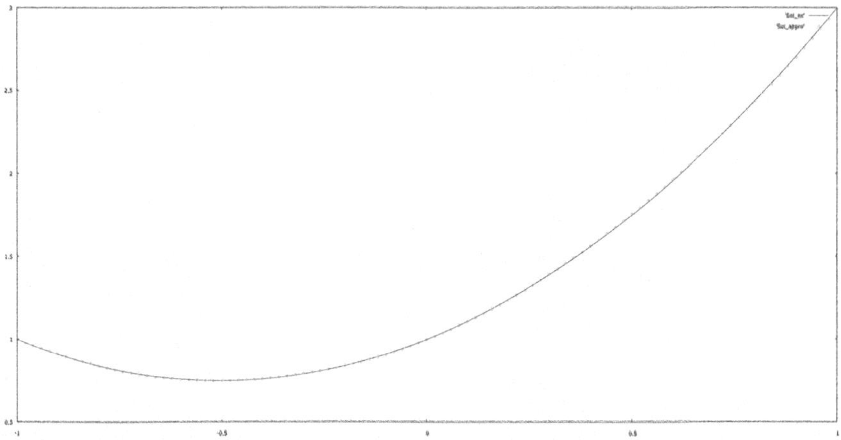

Fig. 5 Exact and computed solutions on Γ_u for Example 1

Example 2 In the second example, we consider data for which the Cauchy problem has an exact solution given by $u(x, y) = \frac{1}{2}(\cos(2xy) + \cos(2x))$.

The results for Example 2 are depicted in Figs. 6 and 7. As in the first example, the algorithm yields a highly accurate solution with an error on the order of 4×10^{-3} away from the inaccessible boundary. There is a slight loss of precision near Γ_u, but the maximum error remains under 2×10^{-3}.

Figure 7 shows that the curves representing the traces on Γ_u of the exact and computed solutions are very close.

The convergence results are consistent with those observed in Example 1, and we do not present them here to avoid overloading the paper.

6 Conclusion

We presented a novel method for addressing the inverse problem associated with the modified biharmonic equation.

This method relies on the factorization of the Cauchy problem for the modified biharmonic equation into two sequential Cauchy problems for Helmholtz and Poisson equations. The factorization is achieved by introducing an auxiliary function, which is utilized to create a sequence of iterates that converge to the solution of the original problem. Each Cauchy problem is solved using a unified algorithm derived from the JN relaxed alternating iterative algorithm [24].

These numerical experiments suggest that the relaxed alternating iterative method is an effective tool for solving this type of ill-posed inverse problem, with the ability to tackle a broad spectrum of such issues.

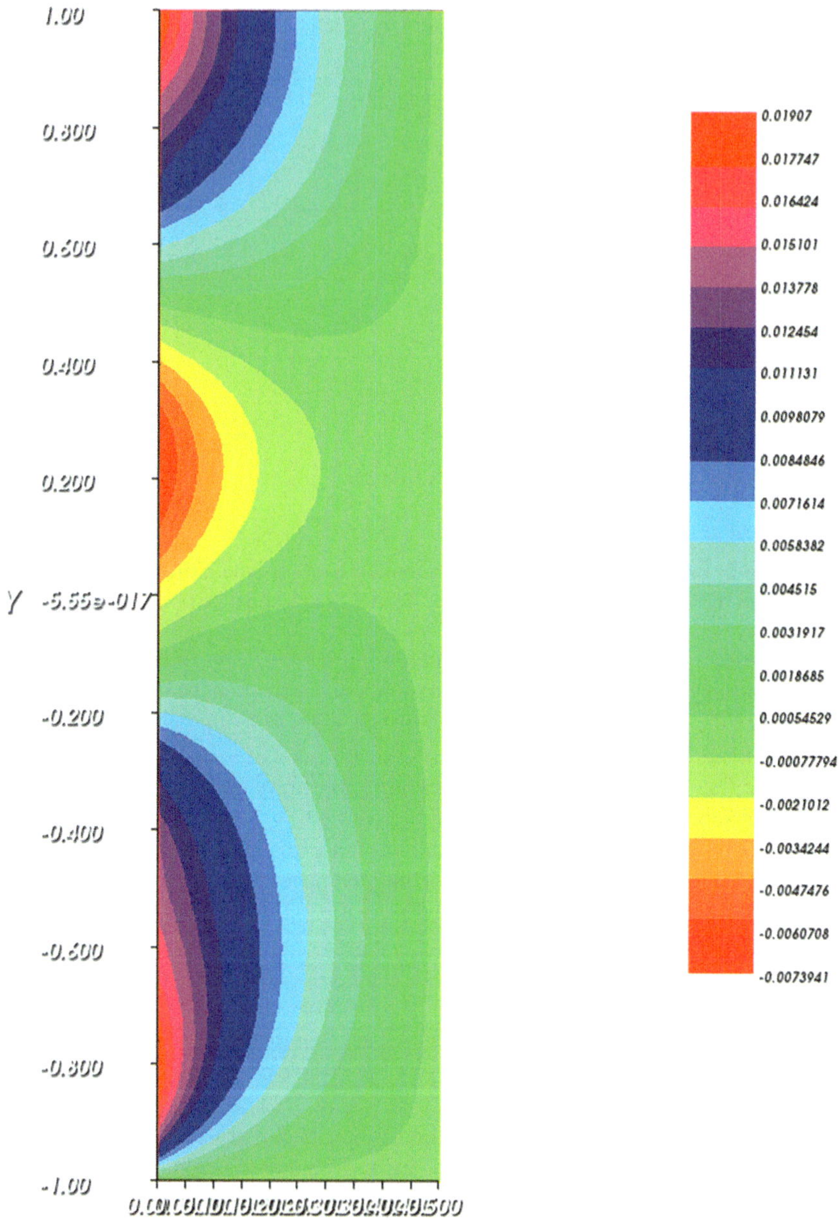

Fig. 6 Error for Example 2

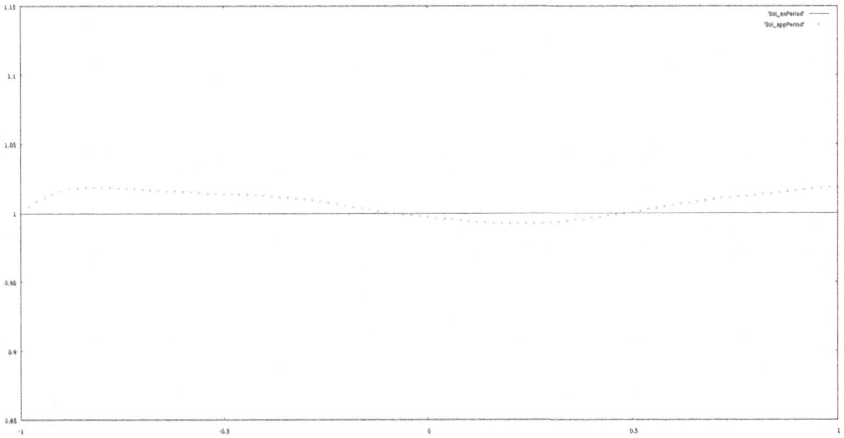

Fig. 7 Exact and computed solutions on Γ_u for Example 2

It is important to note that while the algorithm is relatively straightforward to implement, its convergence properties are highly sensitive to the selection of the relaxation parameter for the unknown boundary condition. In practice, experimenting with different parameters and initial guesses is often necessary to achieve satisfactory results. A careful selection of the relaxation parameter θ to manage the level of regularization and ensure convergence to a stable and accurate solution will likely make this method a popular choice for various applications.

References

1. Aboud, F., Th Jameel, I., Hasan, A.F., Kh Mostafa, B., Nachaoui, A.: Polynomial approximation of an inverse Cauchy problem for Helmholtz type equations. Adv. Math. Models Appl. **7**(3), 306–322 (2022)
2. Aboud, F., Nachaoui, A., Nachaoui, M.: On the approximation of a Cauchy problem in a non-homogeneous medium. J. Phys. Conf. Ser. **1743**(1), 012003 (2021)
3. Afraites, L., Masnaoui, C., Nachaoui, M.: Shape optimization method for an inverse geometric source problem and stability at critical shape. Discrete Contin. Dyn. Syst. Ser. S **15**(1), 1–21 (2022)
4. Arsenashvili, A., Nachaoui, A., Tadumadze, T.: On approximate solution of an inverse problem for linear delay differential equations. Bull. Georgian Natl. Acad. Sci. (N.S.) **2**(2), 24–28 (2008)
5. Benrabah, A., Boussetila, N., Rebbani, F.: Modified auxiliary boundary conditions method for an ill-posed problem for the homogeneous biharmonic equation. Math. Methods Appl. Sci. **43**(1), 358–383 (2020)
6. Berdawood, K., Nachaoui, A., Nachaoui, M.: An accelerated alternating iterative algorithm for data completion problems connected with Helmholtz equation. Stat. Optim. Inf. Comput. **11**(1), 2–21 (2023)
7. Berdawood, K.A., Nachaoui, A., Nachaoui, M., Aboud, F.: An effective relaxed alternating procedure for Cauchy problem connected with Helmholtz equation. Numer. Methods Partial Diff. Equ. **39**(3), 1888–1914 (2022)

8. Boffi, D., Brezzi, F., Fortin, M.: Mixed Finite Element Methods and Applications. Springer Series in Computational Mathematics, vol. 44. Springer, Heidelberg (2013)
9. Chakib, A., Ellabib, A., Nachaoui, A., Nachaoui, M.: A shape optimization formulation of weld pool determination. Appl. Math. Lett. **25**(3), 374–379 (2012)
10. Ciarlet, P.G., Raviart, P.A.: A mixed finite element method for the biharmonic equation. In: Proceedings of the Symposium Mathematical Aspects of Finite Elements in Partial Differential Equations, vol. 33, pp. 125–145. Mathematics Research Center, University of Wisconsin, Madison, Wis. (1974)
11. Ellabib, A., Nachaoui, A.: On the numerical solution of a free boundary identification problem. Inverse Probl. Eng. **9**(3), 235–260 (2001)
12. Ellabib, A., Nachaoui, A.: An iterative approach to the solution of an inverse problem in linear elasticity. Math. Comput. Simul. **77**(2–3), 189–201 (2008)
13. Ellabib, A., Nachaoui, A., Ousaadane, A.: Mathematical analysis and simulation of fixed point formulation of Cauchy problem in linear elasticity. Math. Comput. Simul. **187**, 231–247 (2021)
14. Ellabib, A., Nachaoui, A., Ousaadane, A.: Convergence study and regularizing property of a modified robin-robin method for the Cauchy problem in linear elasticity. Inverse Probl. **38**(7), 075007 (2022)
15. Ellabib, A., Nachaoui, A., Ousaadane, A.: A relaxed robin-robin algorithm for solving a Cauchy problem in linear elasticity. J. Appl. Math. Comput. **71**, 1721–1741 (2025)
16. Essaouini, M., Nachaoui, A., El Hajji, S.: Numerical method for solving a class of nonlinear elliptic inverse problems. J. Comput. Appl. Math. **162**(1), 165–181 (2004)
17. Essaouini, M., Nachaoui, A., El Hajji, S.: Reconstruction of boundary data for a class of nonlinear inverse problems. J. Inverse Ill-Posed Probl. **12**(4), 369–385 (2004)
18. Filippo, G., Christoph, G.H., Guido, S.: Polyharmonic Boundary Value Problems: A Monograph on Positivity Preserving and Nonlinear Higher Order Elliptic Equations in Bounded Domains. Lecture Notes in Mathematics, vol. 1991. Springer-Verlag (2010)
19. Fu, C.L., Feng, X.L., Qian, Z.: The Fourier regularization for solving the Cauchy problem for the Helmholtz equation. Appl. Numer. Math. **59**(10), 2625–2640 (2009)
20. Grunau, Hans-Christoph., Miyake, Nobuhito, Okabe, Shinya: Positivity of solutions to the Cauchy problem for linear and semilinear biharmonic heat equations. Adv. Nonlinear Anal. **10**(1), 353–370 (2021)
21. Hua, Q., Gu, Y., Qu, W., Chen, W., Zhang, C.: A meshless generalized finite difference method for inverse Cauchy problems associated with three-dimensional inhomogeneous Helmholtz-type equations. Eng. Anal. Bound. Elem. **82**, 162–171 (2017)
22. Huang, C.H., Chen, W.C.: A three-dimensional inverse forced convection problem in estimating surface heat flux by conjugate gradient method. Int. J. Heat Mass Transf. **43**(17), 3171–3181 (2000)
23. Jiang, S., Kropinski, M.C.A., Quaife, B.D.: Second kind integral equation formulation for the modified biharmonic equation and its applications. J. Comput. Phys. **249**, 113–126 (2013)
24. Jourhmane, M., Nachaoui, A.: An alternating method for an inverse Cauchy problem. Numer. Algorithm **21**(1), 247–260 (1999)
25. Jourhmane, M., Nachaoui, A.: Convergence of an alternating method to solve the Cauchy problem for Poisson's equation. Appl. Anal. **81**(5), 1065–1083 (2002)
26. Kabanikhin, S.I.: Inverse and Ill-Posed Problems. Walter de Gruyter GmbH & Co. KG, Berlin (2012)
27. Kirsch, A., et al.: An introduction to the mathematical theory of inverse problems. Appl. Math. Sci. **120**. Springer (2011)
28. Lavrentiev, M.M.: Some Improperly Posed Problems of Mathematical Physics. Springer Science & Business Media (2013)
29. Micheletti, A.M., Pistoia, A.: Multiplicity results for a fourth-order semilinear elliptic problem. Nonlinear Anal. Theory Methods Appl. **31**(7), 895–908 (1998)
30. Mukanova, B.: Numerical reconstruction of unknown boundary data in the Cauchy problem for Laplace's equation. Inverse Probl. Sci. Eng. **21**(8), 1255–1267 (2013)

31. Nachaoui, A.: An improved implementation of an iterative method in boundary identification problems. Numer. Algorithms **33**(1–4), 381–398 (2003)
32. Nachaoui, A.: Numerical linear algebra for reconstruction inverse problems. J. Comput. Appl. Math. **162**, 147–164 (2004)
33. Nachaoui, A.: Iterative methods for inverse problems subject to the convection-diffusion equation. Adv. Math. Models Appl. **8**, 327–338 (2023)
34. Nachaoui, A., Aboud, F., Nachaoui, M.: Acceleration of the KMF algorithm convergence to solve the Cauchy problem for Poisson's equation. Springer Proceedings in Mathematics & Statistics, vol. 372, pp. 43–57. Springer Cham (2021)
35. Nachaoui, A., Gontier, L.: Analytical Solutions for Inverse Cauchy Problems: A Comparative Study. Springer Proceedings in Mathematics & Statistics, vol. 498. Springer, Cham (2025)
36. Nachaoui, A., Nachaoui, M.: An hybrid finite element method for a quasi-variational inequality modeling a semiconductor. RAIRO Oper. Res. **57**(4), 2197–2207 (2023)
37. Nachaoui, A., Nachaoui, M., Chakib, A., Hilal, M.: Some novel numerical techniques for an inverse Cauchy problem. J. Comput. Appl. Math. **381**, 113030 (2021)
38. Nachaoui, A., Nachaoui, M., Tadumadze, T.: Meshless Methods to Noninvasively Calculate Neurocortical Potentials from Potentials Measured at the Scalp Surface. Springer Proceedings in Mathematics & Statistics, vol. 428, pp. 99–117. Springer Cham (2023)
39. Nachaoui, A., Rashid, S.M.: A Mesh Free Wavelet Method to Solve the Cauchy Problem for the Helmholtz Equation. Springer Proceedings in Mathematics & Statistics, vol. 428, pp. 77–98. Springer Cham (2023)
40. Nachaoui, A., Salih, H.W.: An analytical solution for the nonlinear inverse Cauchy problem. Adv. Math. Models Appl. **6**(3), 191–206 (2021)
41. Nachaoui, A., Rasheed, S.M., Sadiq, G.W.: A Semi-analytical Solution for Inverse Cauchy Problems. Springer Proceedings in Mathematics & Statistics, vol. 498. Springer, Cham (2025)
42. Nachaoui, A., Sadiq, G.W.: Identification of a Space Dependent Force in a Hyperbolic Equation using a Polynomial Expansion. Springer Proceedings in Mathematics & Statistics, vol. 498. Springer, Cham (2025)
43. Nachaoui, A., Sadiq, G.W.: Investigation of an inverse problem in heat propagation. To appear (2025)
44. Nachaoui, M., Nachaoui, A., Tadumadze, T.: On the numerical approximation of some inverse problems governed by nonlinear delay differential equation. RAIRO Oper. Res. **56**(3), 1553–1569 (2022)
45. Qian, A.L., Xiong, X.T., Wu, Y.-J.: On a quasi-reversibility regularization method for a Cauchy problem of the Helmholtz equation. J. Comput. Appl. Math. **233**(8), 1969–1979 (2010)
46. Qian, Z., Feng, X.: A fractional Tikhonov method for solving a Cauchy problem of Helmholtz equation. Appl. Anal. **96**, 1656–1668 (2017)
47. Rashid, S.M., Nachaoui, A.: A Haar wavelets-based direct reconstruction method for the Cauchy problem of the Poisson equation. Discrete Contin. Dyn. Syst. Ser. S **18**(1), 148–168 (2025)
48. Reddy, G.M.M., Nanda, P., Vynnycky, M., Cuminato, J.A.: An adaptive boundary algorithm for the reconstruction of boundary and initial data using the method of fundamental solutions for the inverse Cauchy-Stefan problem. Comput. Appl. Math. **40**(3) (2021)
49. Regińska, T., Regiński, K.: Approximate solution of a Cauchy problem for the Helmholtz equation. Inverse Prob. **22**(3), 975–989 (2006)
50. Shi, R., Wei, T., Qin, H.H.: A fourth-order modified method for the Cauchy problem of the modified Helmholtz equation. Numer. Math. Theory Methods Appl. **2**(3), 326–340 (2009)
51. Yarmukhamedov, Sh., Yarmukhamedov, I.: Cauchy problem for the Helmholtz equation. In: Ill-posed and Non-classical Problems of Mathematical Physics and Analysis, pp. 143–172. VSP, Utrecht, Inverse Ill-posed Probability Series (2003)

A Hybrid Image Restoration Approach: Integrating Coherence Transport and Anisotropic Diffusion

Mohamed Nacereddine Toros, Mourad Nachaoui, Mustapha Johri, and Amine Laghrib

Abstract Image inpainting is a crucial technique in image processing, focused on recovering degraded images by reducing noise and reconstructing missing regions. In this paper, we propose a hybrid approach that integrates coherence transport and anisotropic diffusion to enhance the quality of restoration. Our method utilizes a coupled equation, where coherence transport facilitates smooth transitions across the image, while anisotropic diffusion helps preserve important image structures. This combination effectively balances edge preservation and uniform diffusion, yielding high-quality restored images with improved visual fidelity.

Keywords Digital image inpainting · Laplace interpolation · Linear elliptic · Direct problem · Fluid dynamics · Coherence transport · Homogeneous diffusion

1 Introduction

In the last few years, a number of PDE-based inpainting models have appeared. Some models are written down directly, based on some mathematical understandings on the properties of diffusion PDEs (the heat equation, the Cahn-Hilliard equation, etc.) or in analogy with linear and nonlinear filtering [1–4], by the construction of gradient

---

Mustapha Johri and Amine Laghrib are the authors contributed equally to this work.

M. N. Toros (✉) · M. Nachaoui (✉)
EMI, Sultan Moulay Slimane University, Beni Mellal, Morocco
e-mail: mohamed_nacereddine.toros@usms.ac.ma

M. Nachaoui
e-mail: m.nachaoui@usms.ma

M. Johri
Department of Applied Mathematics and Informatics, ENSAM, Mohammed V University, Rabat, Morocco
e-mail: m.johri@um5r.ac.ma

A. Laghrib
LMA ENSA Khouribga, Sultan Moulay Slimane University, Beni Mellal, Morocco

A. Laghrib et al. (eds.), *Computational Methods for Inverse Problems and Applications*, Springer Proceedings in Mathematics & Statistics 498,
https://doi.org/10.1007/978-3-031-89498-5_11

flows [5–7] or by active contours modeling [8–10]. Others are formulated as an adequate optimization problem. We better explain in the next section PDE-based approaches and we give an overview on them.

The underlying idea of PDE-based methods is to interpolate image data across all boundaries of the missing area D via a process modeled by the following k-th-order evolutionary PDE:

$$\partial_t u - \mathcal{R}(x, u, Du, \ldots, D^k u) = 0, \quad \text{in } D, \quad u = f, \quad \text{on } \partial D, \tag{1}$$

where $\mathcal{R}$ is a k-th differential operator. The choice of the operator $\mathcal{R}$ is inspired by phenomenological modeling like Navier-Stokes and fluid dynamics in [11–13], or inspired from phase field models like the Cahn-Hilliard equation [14–17]. In other instances, it comes from the idea of propagating local information with smoothness constraints, in analogy with physical phenomena like heat propagation in physical structures by a diffusion process. These diffusion PDEs take their origin from the following general continuity equation:

$$\partial_t u = \nabla \cdot j, \tag{2}$$

where j is expressed by Fick's law as follows:

$$j = -D\nabla u, \tag{3}$$

and D is a diffusion tensor intended to be a positive definite symmetric matrix which encodes the properties of the diffusion. In cases where j and ∇u are parallel, equation (1.13) is called isotropic; we may replace the diffusion tensor by a positive scalar-valued diffusivity. In general, i.e., where j and ∇u are not parallel, the equation is called anisotropic. Various investigations are focusing on the choice of the diffusion tensor D: a positive diffusion tensor $D(x, y)$, i.e., depends only on z and not on the image u, leads to linear, isotropic or anisotropic PDEs which are the simplest PDE methods and mostly used for image smoothing. However, over time, other strong character models have emerged. These exploits in image inpainting because they reduce blur and smooth edges. Then, other alternatives are focusing on adaptive nonlinear diffusion methods which are especially oriented towards understanding the mathematical properties of the diffusion and to drive it by choosing D depending on specific image characteristics.

The second-order PDEs have the advantage of being well-established theoretically and easy to solve numerically. However, they do not usually perform well in inpainting, chiefly when the size of the hole in the image is important. In fact, the major drawbacks are either the disconnection of level lines over large distances (Connectivity Principle), either their smooth propagation into the damaged domain, or the inability to reproduce features of higher order (curvature, corners, etc.) due to the lack of information (second-order). Let us now start with the presentation of some existing second-order inpainting models.

Also the aims of the Smooth Inpainting via Harmonic Extension [18] is the Harmonic extension of the known image f on $\Omega \setminus D$ by a function u which solves the following PDE:

$$\Delta u = 0, \quad \text{in } D \quad \text{and} \quad u|_{\partial D} = f, \tag{4}$$

The shortcomings of second-order methods gave rise to a new class of higher-order diffusion models which generally perform better. In fact, whenever the image is contaminated by noise, they dampen oscillations and high frequencies (noise) faster than second-order-based diffusion models. In addition, they are richer than second-order ones and have the benefits of connecting isolines across missing parts while preserving the curvature (of third-order). Moreover, the use of boundary conditions for both the solution $u(x)$ and its derivatives provides supplementary information on the isoline directions and allows matching edges across large distances.

Several contributions focus on higher-order PDEs by incorporating higher-order derivatives or a sophisticated combination of higher- and second-order ones.

Bertalmio Approach and Fluid Interpretation the pioneering work within PDE-based approaches is that of Bertalmio et al. [12], where the authors proposed to propagate both the gradient direction (geometry) and the grayscale values of the image inside the region to be filled by solving the PDE:

$$\partial_t u - \nabla^{\perp} \cdot \nabla \Delta u = 0, \quad \text{in } D, \quad u = f, \quad \text{on } \partial D, \tag{5}$$

where $\nabla^{\perp}$ denotes the perpendicular gradient $(-\partial_y, \partial_x)$, and Δ is the Laplace operator. In [11], Bertalmio et al. exploited ideas from classical computational fluid dynamics in the image inpainting problem. Their approach is directly based on the Navier-Stokes equations for fluid dynamics and is given by the following equation:

$$w = \Delta u, \quad \partial_t w + v \cdot \nabla w = \sigma \nabla \cdot (g(\|\nabla w\|)\nabla w), \tag{6}$$

where the velocity field $v = \nabla^{\perp} u$ and the function σ allows for anisotropic diffusion of the smoothness w. In analogy with fluid dynamics, the image intensity is taken as a "stream function" for two-dimensional incompressible flow, and its Laplacian plays the role of the vorticity of the fluid.

The mogeneous Diffusion [19] uses the constant diffusivity $g = 1$. This leads to the linear diffusion equation (also known as the heat equation) $\frac{\partial u}{\partial t} = \Delta u$ with the Laplacian $\Delta = \partial_{xx} + \partial_{yy}$. This evolution generates the well-known Gaussian scale-space [19]. It is simple to implement and uses no additional parameters apart from t. Since homogeneous diffusion smooths equally at all locations of the image, it is space-invariant. However, as it cannot distinguish between noise and edges, it also blurs semantically important image edges.

Let $\Omega \subset \mathbb{R}^d$ (where $d = 2, 3$) denote the entire image domain. The problem is to fill in image information in the incomplete or damaged region $D \subset \Omega$ based on the image information available outside D (i.e., in $\Omega \setminus D$). We assume that ∂D is sufficiently smooth. When information is available near the entire boundary ∂D, it can be used as Dirichlet boundary conditions for the PDE that propagates the information inside D as in [18, 20, 21]. The inpainting problem is formulated as a Cauchy problem, which is linear for smooth images and nonlinear for images containing edges. In both cases, we consider the following inverse boundary-value problem for the unknown image function u:

$$\begin{cases} \nabla \cdot \left[k\left(|\nabla u|^2\right) \nabla u\right] = 0, & \text{in } D, \\ u = f, & \text{on } \partial D, \\ k\left(|\nabla u|^2\right) \nabla u \cdot n = \psi, & \text{on } \partial D, \end{cases} \tag{7}$$

where n is the unit outward normal to ∂D, $f \in H^{1/2}(\partial D)$, and $\psi \in \left(H_{00}^{1/2}(\partial D)\right)'$ are given functions. Here, $\left(H_{00}^{1/2}(\partial D)\right)'$ denotes the dual space of $H_{00}^{1/2}(\partial D)$.

We assume the following conditions on the function $k(\cdot)$:

$$\begin{cases} \text{(a)} \quad \beta_1 \le k(r) \le \beta_0 > 0, & \forall r \in \mathbb{R}^+, \\ \text{(b)} \quad k'(r) \le 0, & \forall r \in \mathbb{R}^+, \\ \text{(c)} \quad k(r) + 2rk'(r) \ge \beta_2 > 0, & \forall r \in \mathbb{R}^+, \\ \text{(d)} \quad \beta_3 > 0, \quad (B(\omega)X, X) \ge \beta_3|X|^2, & \forall X \in \mathbb{R}^d, \end{cases} \tag{8}$$

where $B(\omega) = k(|\omega|^2)I_d + 2k'(|\omega|^2)\omega \otimes \omega$, $\omega \in \mathbb{R}^d$, and I_d is the identity matrix in $\mathbb{R}^d$.

We note that the linear case corresponds to $k(\cdot) \equiv 1$, which satisfies all the conditions listed above.

In linear case where $k(r) \equiv 1$, we have a Cauchy problem for the Laplace operator. A Nash game formulation of this problem was studied in [22, 23]. The cost functionals are elliptic and convex. The authors proved that there always exists a unique Nash equilibrium, which corresponds to the reconstructed data when the Cauchy problem has a solution.

2 The Proposed Smoothness-Anisotropic System

Let $\Omega \subset \mathbb{R}^d$ (with $d = 2, 3$) denote the entire image domain, which has a piecewise smooth boundary $\partial\Omega$. Let $D \subset \Omega$ be a damaged subregion where the information is missing. A PDE-based inpainting model aims to fill in image information in the incomplete region D using the available information outside D, i.e., in $\Omega \setminus D$.

In recent years, numerous PDE-based inpainting models have been proposed. Some models are directly formulated based on mathematical insights into diffusion PDE properties, such as the heat equation and the Cahn-Hilliard equation. Others are derived as minimizations of appropriately regularized functionals. In [18], Chan and Shen extended the total variation (TV) based denoising model of Rudin, Osher, and Fatemi [24] for inpainting purposes. In [25], they proposed the Curvature Driven Diffusion (CDD) model, where the diffusion coefficient depends on the curvature

Fig. 1 Direct problem case

of the isophotes. Another approach to inpainting is based on the complex Ginzburg-Landau equation [26].

In our work, we consider the following diffusion function:

$$k_\alpha(r) = \frac{1}{\sqrt{r + \varepsilon^2}} + \alpha, \quad \forall r \in \mathbb{R}^+,$$

where $\alpha > 0$ and $\varepsilon > 0$ are small parameters. The function $k_\alpha(\cdot)$ is a regularized version of the diffusion function used in the TV model [18], which aims to join the same level lines on both sides of D and minimize their lengths. In the standard TV model, i.e., without the regularizing parameter α, the solution must lie in $W^{1,1}(D)$. The existence and uniqueness of solutions for the mixed direct problem were studied in [27]. However, no results are currently available for Cauchy problems in the $W^{1,1}(D)$ space. To work in the Hilbert space $H^1(D)$, where the Cauchy problem has been theoretically and numerically studied, we add the constant α in $k_\alpha(\cdot)$. This choice is compatible with the natural conditions in (2), ensuring the existence and uniqueness of solutions for the direct and Cauchy problems.

The region D is entirely within Ω (i.e., $\partial D \cap \partial\Omega = \emptyset$), as illustrated in Fig. 1. Here, the information available near the boundary ∂D can be used as Dirichlet boundary conditions for the PDE that propagates information inside D. The problem then becomes:

$$\begin{cases} \nabla \cdot \left[k_\alpha(|\nabla u|^2)\nabla u\right] = 0, & \text{in } D, \\ u = f, & \text{on } \partial D. \end{cases} \tag{9}$$

The goal is to interpolate image data across the boundaries of the missing area (D). This process is modeled by solving the following fourth-order evolutionary partial differential equation (PDE) for transport inpainting:

$$u_t = \nabla^\perp u \cdot \nabla \Delta u,$$

which convects the image intensity u along level lines Δu, chosen as a smoothness estimator of the image function.

Our approach involves a hybrid partial differential equation (PDE) that integrates the strengths of coherence transport and anisotropic diffusion. The combined equation is given by:

$$u_t = \eta \underbrace{\nabla^{\perp} u \cdot \nabla \Delta u}_{\text{Smoothness transport}} + \nu \underbrace{\nabla \cdot (k_\alpha(\|\nabla u\|)\nabla u)}_{\text{Anisotropic diffusion}}. \tag{10}$$

where $\nu > 0$ is a weight anisotropic diffusion parameter and k_α is a monotonic decreasing function of the image gradient with the regularization parameter (α)

The term $\eta\nabla^{\perp}u \cdot \nabla\Delta u$ represents the smoothness transport component, which aligns with coherence transport principles. This term transports smoothness along the flow of image structures, helping to reduce noise while preserving the directionality of edges and textures.

The term $\nu\nabla \cdot (k_\alpha(\|\nabla u\|)\nabla u)$ corresponds to anisotropic diffusion, where $k_\alpha(\|\nabla u\|)$ is a function of the gradient magnitude $\|\nabla u\|$. This function controls the amount of diffusion, allowing strong smoothing in homogeneous regions while preserving edges where the gradient is high. For the linear case k_α= 1, we have a Cauchy problem for the Laplace operator. and our system can be like:

$$\begin{cases} \eta\nabla^{\perp}u \cdot \nabla\Delta u + \nu\Delta u = 0, & \text{in } D, \\ u = f, & \text{on } \partial D. \end{cases} \tag{11}$$

3 Discretization Scheme

Consider the following PDE:

$$u_t = \eta\nabla^{\perp}u \cdot \nabla\Delta u + \nu\nabla \cdot (k_\alpha(\|\nabla u\|)\nabla u).$$

We discretize this equation using finite differences as follows:

- Let u_i^n represent the value of u at the spatial grid point i and at time step n.
- Δt is the time step, and Δx is the space step.
- The update equation is given by:

$$u_i^{n+1} = u_i^n + \Delta t \left[\eta\nabla^{\perp}u_i^n \cdot \nabla\Delta u_i^n + \nu\nabla \cdot \left(k_\alpha(\|\nabla u_i^n\|)\nabla u_i^n\right)\right].$$

3.1 Time Discretization

The time derivative is discretized using a forward Euler scheme:

$$\frac{u_i^{n+1} - u_i^n}{\Delta t} = u_t.$$

Thus, the update rule becomes:

$$u_i^{n+1} = u_i^n + \Delta t\, f(u_i^n, \nabla u_i^n, \nabla^2 u_i^n),$$

where $f(u_i^n, \nabla u_i^n, \nabla^2 u_i^n)$ represents the right-hand side of the PDE.

3.2 Space Discretization

- The gradient and Laplacian of u are discretized using central differences:

$$\nabla u_i \approx \frac{u_{i+1} - u_{i-1}}{2\Delta x}, \quad \nabla^2 u_i \approx \frac{u_{i+1} - 2u_i + u_{i-1}}{\Delta x^2}.$$

- The smoothness transport term becomes:

$$\nabla^{\perp} u_i^n \cdot \nabla \Delta u_i^n \approx -\frac{\partial_y u}{\Delta x} \cdot \frac{\partial_x^2 u}{\Delta y} + \frac{\partial_x u}{\Delta y} \cdot \frac{\partial_y^2 u}{\Delta x}.$$

- The anisotropic diffusion term becomes:

$$\nabla \cdot (k_\alpha(\|\nabla u\|)\nabla u) \approx \frac{1}{\Delta x}\left(k_\alpha(\|\nabla u\|_{i+1})\frac{u_{i+1} - u_i}{\Delta x} - k_\alpha(\|\nabla u\|_{i-1})\frac{u_i - u_{i-1}}{\Delta x}\right).$$

Finally, we obtain the following Algorithm to resolve our main problem.

4 Numerical Results

In this section, we present the numerical results obtained using our approach described in the previous section. To assess the effectiveness of the proposed approach, we conducted a numerical study on various images. In all numerical experiments, we selected images where the missing region D satisfies $\partial D \cap \partial\Omega \neq \emptyset$, as shown in Fig. 1. In each example,the damaged regions are highlighted in black. We also provide numerical results that demonstrate the stability of our method with respect to noisy data.

We recall that the Mean Square Error (MSE) and Peak Signal-to-Noise Ratio (PSNR) are used to evaluate the quality of an image I_2 with respect to a reference image I_1. These metrics are defined as follows:

$$\text{MSE} = \frac{1}{mn}\sum_{i=0}^{m-1}\sum_{j=0}^{n-1} |I_1(i,j) - I_2(i,j)|^2,$$

Algorithm 1 Finite Difference Scheme for Solving the PDE

1: **Input:** Initial condition $u_i^0 = f$ for all grid points i, time step Δt, space step Δx, and final time T
2: **Initialize:** u_i^0 for all grid points i
3: **for** each time step n such that $t_n \leq T$ **do**
4: **for** each grid point i **do**
5: Compute the smoothness transport term:

$$\nabla^{\perp} u_i^n \cdot \nabla \Delta u_i^n$$

6: Compute the anisotropic diffusion term:

$$\nabla \cdot \left(k_\alpha(\|\nabla u_i^n\|)\nabla u_i^n\right)$$

7: Update u_i^{n+1} using:

$$u_i^{n+1} = u_i^n + \Delta t\left[\eta \nabla^{\perp} u_i^n \cdot \nabla \Delta u_i^n + \nu \nabla \cdot \left(k_\alpha(\|\nabla u_i^n\|)\nabla u_i^n\right)\right]$$

8: **end for**
9: **end for**
10: **Output:** Final solution u_i^N at time t_N

$$\text{PSNR} = 10 \cdot \log_{10}\left(\frac{\text{MAX}_{I_2}^2}{\text{MSE}}\right),$$

where m and n denote the number of rows and columns of the matrix representing the image, and MAX_{I_2} is the maximum possible pixel value in the image I_2.

We begin by applying the proposed model to the problem of image inpainting, with a focus on cases where the missing regions do not contain edge information. Our experiments illustrate the use of this method for color image restoration. A color image can be represented by three separate channels: u_R, u_G, and u_B, corresponding to the red, green, and blue color intensities, respectively. This extension to color images follows naturally from the framework used for grayscale images, where the inpainting process is performed independently for each channel by solving a Cauchy problem for each one.

We first focus on the *Island* image, where a portion of the image is artificially damaged using a rectangular mask as presented in Fig. 2. Our goal is to recover the missing region using different scenarios based on the proposed algorithm. For each scenario, we explore various inpainting strategies, evaluating the algorithm's effectiveness in reconstructing the damaged areas. These experiments help assess the algorithm's ability to preserve image structure and color fidelity under different conditions of missing data. By comparing the results across scenarios, we gain insights into the strengths and limitations of the proposed approach in handling real-world image inpainting tasks.

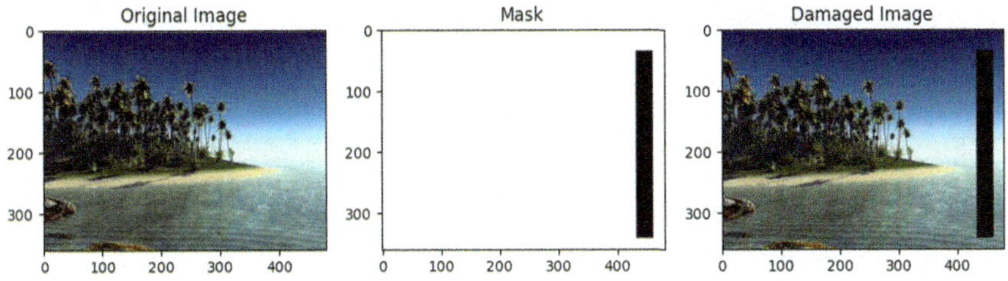

Fig. 2 The construction of the simulated *Island* image using a rectangular mask

4.1 *Case 1:* $(\eta, \nu) = (1, 0)$

In the first scenario, we evaluate the performance of the proposed algorithm when only the regularization term is active, and no data fidelity term is included. This setup isolates the effect of the regularization term, which controls the smoothness of the inpainted region without relying on the known values of the missing pixels. This is the case when $(\eta, \nu) = (1, 0)$.

Figure 3 present the images in three stages: the original image, the noisy image (with missing data), and the image restored by the algorithm under this regularization-only scenario.

As expected, the regularization term significantly smooths the image, reducing high-frequency noise in the damaged regions. This is especially evident in areas where the original image contained small details that are lost or blurred due to the smoothing effect. While the regularization term helps to recover smooth part, it also has the downside of blurring sharp edges, which reduces fine details. One of the trade-offs in this scenario is the preservation of edges. Since no data fidelity is enforced, the algorithm attempts to fill in the missing regions by relying solely on the regularization term, which typically leads to the loss of sharp boundaries. In cases where edges are less pronounced in the noisy image, the inpainting process can lead to smoother transitions, but at the cost of edge accuracy.

The regularization-only scenario demonstrates the primary role of the smoothness prior in image inpainting: noise reduction and global structure preservation. However,

(a) Original (b) Recovered (c) Error

Fig. 3 The recovered *Island* image using the proposed Algorithm where $(\eta, \nu) = (1, 0)$

(a) Original (b) Recovered (c) Error

Fig. 4 The recovered *Island* image using the proposed Algorithm where $(\eta, \nu) = (0, 0.1)$

the absence of data fidelity leads to significant degradation in edge sharpness and a noticeable loss of detail, as quantified by the MSE, PSNR, and SSIM metrics. While the image is smoothed and appears less noisy, the trade-off is the loss of high-frequency components and fine-grained details. This scenario illustrates the importance of incorporating both regularization and data fidelity in real-world image restoration tasks, where edge preservation is often crucial.

4.2 *Case 2:* $(\eta, \nu) = (0, 1)$

Present results where only the data fidelity term is active (no regularization). Explain how this affects noise and edge preservation. For the second scenario, Fig. 4 presents the *Island*: the original image, and the image restored by the algorithm under the regularization-only scenario. In this setup, the algorithm aims to recover the image by applying a regularization technique, without additional constraints or data enhancement. As shown in the figure, the original image serves as the baseline, while the noisy image represents the degradation caused by missing information or noise. The restored image, obtained after applying the regularization method, exhibits a partial recovery of the original details. However, it is evident that the restoration process has not fully eliminated the noise or reconstructed the missing data, leading to visible artifacts or blur in certain areas. This highlights the limitations of using regularization alone in scenarios where data is missing or severely degraded.

By comparing the result with the original image, we can detect some blur left in the recovered surface. This residual blur suggests that the recovery process has not fully restored the image to its original sharpness, due to limitations in the impainting algorithm.

(a) Original (b) Recovered (c) Error

Fig. 5 The recovered *Island* image using the proposed Algorithm where $(\eta, \nu) = (0.5, 0.5)$

4.3 *Case 3:* $(\eta, \nu) = (0.5, 0.5)$

Present results where both terms are active with the same controlled parameters. For the third scenario, Fig. 5 presents the Island: the original image and the image restored by the algorithm under the presence of both regularization and data enhancement constraints. In this setup, the algorithm aims to recover the image by applying a regularization technique alongside additional constraints focused on improving data enhancement. The regularization term helps to preserve the overall structure and reduce noise, while the data enhancement constraints aim to further refine the image by restoring missing or degraded data. As shown in the figure, the restored image is closer to the original, with sharper details and fewer artifacts compared to the restoration achieved with regularization alone. The combination of both terms allows for a more accurate recovery of the image, making the algorithm more robust to noise and data loss.

4.4 *Comparison Between Scenarios*

To compare the three scenarios in recovering missing data, we consider an example of *Island 2* and we compare the recovered image using the three scenarios in Fig. 6. Specifically, we will use PSNR and SSIM metrics to quantitatively assess the performance of each restoration scenario. In Table 1, we present the PSNR and SSIM values for each of the three scenarios: regularization-only restoration, regularization with data enhancement, and the noisy image as the baseline. This allows us to compare how effectively each method recovers missing data and restores the image to its original quality. This table shows the best restoration outcome in scenarios 3, where combining regularization with data enhancement techniques yields the highest PSNR and SSIM values. This method achieves the most accurate recovery of missing data, with fewer artifacts and better overall quality.

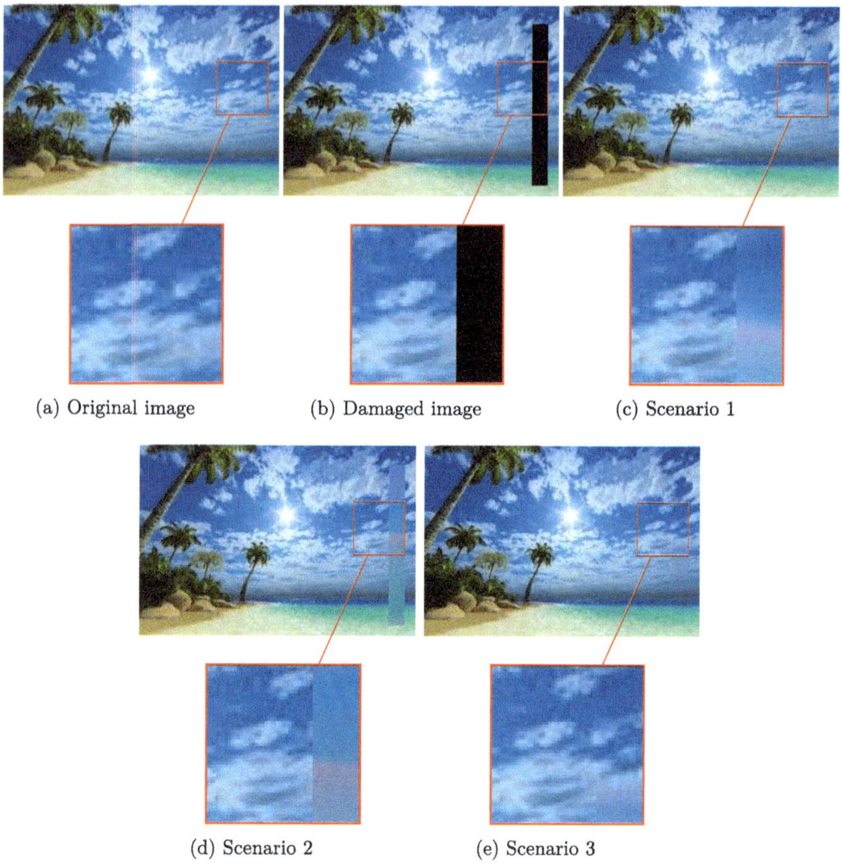

Fig. 6 Comparison between the three scenarios for the (*Island 2* image) recovery

Table 1 PSNR and SSIM comparison for different restoration scenarios on *Island 2*. Higher PSNR and SSIM values indicate better recovery

Scenario	PSNR (dB)	SSIM	Comments
Scenario 1	28.74	0.797	Low quality, significant blur and missing data
Scenario 2	28.91	0.810	Improved but still some artifacts in missing data areas
Scenario 3	30.25	0.822	Best recovery, clearer details and reduced artifacts with small blur

5 Conclusion

We have proposed a novel image restoration method based on a coupled coherence transport-homogeneous diffusion equation. This hybrid approach leverages the benefits of both smoothness transport and anisotropic diffusion, providing a balance between noise reduction and edge preservation. Experimental results demonstrate that our method effectively restores degraded images while maintaining important structural details.

References

1. Nachaoui, M., Afraites, L., Hadri, A., Laghrib, A.: A non-convex non-smooth bi-level parameter learning for impulse and gaussian noise mixture removing. Commun. Pure Appl. Anal. **21**(4), 1249–1291 (2022)
2. Oliveira, M., Bowen, R.M. B., Chang, Y.-S.: Fast digital image inpainting. In: Proceedings of the International Conference on Visualization, Imaging and Image Processing (VIIP 2001), pp. 261–266. ACTA Press (2001)
3. Nachaoui, M., Laghrib, A.: An improved bilevel optimization approach for image super-resolution based on a fractional diffusion tensor. J. Franklin Inst. **359**(13), 7165–7195 (2022)
4. Alahyane, M., Hakim, A., Laghrib, A., Raghay, S.: Fluid image registration using a finite volume scheme of the incompressible Navier-stokes equation. Inverse Probl. Imaging **12**(5) (2018)
5. Alahyane, M., Hakim, A., Laghrib, A., Raghay, S.: A lattice Boltzmann method applied to the fluid image registration. Appl. Math. Comput. **349**, 421–438 (2019)
6. Lekbir, A., Hadri, A., Laghrib, A., Nachaoui, M.: A non-convex denoising model for impulse and gaussian noise mixture removing using bi-level parameter identification. Inverse Probl. Imaging **16**(4), 827–870 (2022)
7. Laghrib, A., Hadri, A., Hakim, A., Raghay, S.: A new multiframe super-resolution based on nonlinear registration and a spatially weighted regularization. Inf. Sci. **493**, 34–56 (2019)
8. Esedoglu, S., Shen, J.: Digital image inpainting by Mumford-Shah-Euler model. Eur. J. Appl. Math. **13**, 353–370 (2002)
9. Tsai, A., Yezzi, A., Willsky, A.S.: Curve evolution implementation of the Mumford-Shah functional for image segmentation, denoising, interpolation, and magnification. IEEE Trans. Image Process. **10**(8), 1169–1186 (2001)
10. Nachaoui, M., Afraites, L., Laghrib, A.: A regularization by denoising super-resolution method based on genetic algorithms. Signal Process. Image Commun. **99**, 116505 (2021)
11. Bertalmio, M., Bertozzi, A.L., Sapiro, G.: Navier-stokes, fluid dynamics, and image and video inpainting. In: Proceedings of the IEEE Computer Vision and Pattern Recognition (CVPR), pp. 355–362 (2001)
12. Bertalmio, M., Sapiro, V.C. G., Ballesteri, C.: Image inpainting. In: Proceedings of the 27th Annual Conference on Computer Graphics and Interactive Techniques, New York, ACM, pp. 417–427 (2000)
13. Laghrib, A., Hadri, A., Hakim, A.: An edge preserving high-order PDE for multiframe image super-resolution. J. Franklin Inst. **356**(11), 5834–5857 (2019)
14. Bertozzi, A., Esedoglu, S., Gillette, A.: Analysis of a two-scale Cahn-Hilliard model for image inpainting. Multiscale Model. Simul. **6**(3), 913–936 (2007)
15. Bertozzi, A., Esedoglu, S., Gillette, A.: Inpainting of binary images using the Cahn-Hilliard equation. IEEE Trans. Image Proc. **16**(1), 285–291 (2007)

16. Bertozzi, A., Schonlieb, C.B.: Unconditionally stable schemes for higher order inpainting. Commun. Math. Sci. **9**(2), 413–457 (2011)
17. Idriss, E.M., Laghrib, A., Hadri, A., et al.: A theoretical study of a bilateral term with a tensor-based fourth-order PDE for image super-resolution. Adv. Comput. Math. **48**(83), 83 (2022). https://doi.org/10.1007/s10444-022-09996-6
18. Chan, T., Shen, J.: Mathematical models for local nontexture in paintings. SIAM J. Appl. Math. **62**(3), 1019–1043 (2002)
19. Iijima, T.: Basic theory on normalization of pattern (in case of typical one-dimensional pattern). Bull. Electrotech. Lab. **26**, 368–388 (1962)
20. Ait Bella, F.Z., El Rhabi, M., Hakim, A., Laghrib, A.: Reduction of the non-uniform illumination using nonlocal variational models for document image analysis. J. Franklin Inst. **355**(16), 8225–8244 (2018). https://doi.org/10.1016/j.jfranklin.2018.08.012
21. Boukdir, A., Laghrib, A., Nachaoui, M.: Time-fractionated anisotropic osmosis with bilateral total variation for effective shadow mitigation in image processing. Appl. Math. Model. **137**, 115671 (2025). https://doi.org/10.1016/j.apm.2024.115671
22. Habbal, A., Kallel, M.: Data completion problems solved as Nash games. J. Phys. Conf. Ser. **386**, 012004 (2012)
23. Habbal, A., Kallel, M.: Neumann-Dirichlet Nash strategies for the solution of elliptic Cauchy problems. SIAM J. Control. Optim. **51**, 4066–4083 (2013)
24. Rudin, L.I., Osher, S., Fatemi, E.: Nonlinear total variation based noise removal algorithms. Phys. D **60**, 258–268 (1992)
25. Chan, T., Shen, J.: Non-texture inpainting by curvature-driven diffusions. J. Visual Commun. Image Represent. **12**, 436–449 (2001)
26. Grossauer, H., Scherzer, O.: Using the complex Ginzburg-Landau equation for digital inpainting in 2D and 3D. In: Griffin, L.D., Lillholm, M. (eds.) Scale Space Methods in Computer Vision, pp. 225–236. Springer, Berlin Heidelberg (2003)
27. Temam, R.: Applications de L'analyse Convexe Au Calcul des Variations. In: Nonlinear Operators and the Calculus of Variations, pp. 208–237. Springer (Summer School, Univ. Libre Bruxelles, Brussels, 1975), Lecture Notes in Math, Berlin (1976)

Total Variation Regularized Sparse Image Deconvolution via Accelerated Condat-Vũ Algorithm

Yasmine El Mobariki and Amine Laghrib

Abstract In the current manuscript, we delineate an Accelerated Condat-Vũ (ACV) algorithm to tackle the complexities associated with the deconvolution problem. In contrast to the traditional Condat-Vũ approach, the ACV framework achieves ideal convergence rates and greatly outperforms the deterministic algorithm, as demonstrated in Driggs in SIAM J Imag Sci 17(4):2076–2109 [1]. Our investigation goes beyond theoretical validation, delving into numerical experiments to illustrate the method's increased efficiency over its predecessors, highlighting its practical applicability in real-world deconvolution circumstances.

Keywords Total variation · Convex optimization · Proximity operator · Denoising · Image deblurring

1 Introduction

Image deconvolution, also known as deblurring, recovers image sharpness lost owing to numerous sources of blur. This technique is essential in sectors with excessive image quality, which include astronomy, microscopy, remote sensing, and graphics [2–14]. However, researchers and physicians working in fields ranging from medical imaging to astronomical observation routinely come across images that have been degraded by various physical processes [15, 16]. A radiologist examining an MRI scan, an astronomer studying distant galaxies, and a microscopist researching cellular structures all face the same venture of extracting significant data from imperfect observations. While image deconvolution can significantly improve image quality, it is important to recognize that the effectiveness of those strategies can range primarily based on the precise software and the sort of noise and blur present in the images [17–19].

These authors contributed equally to this work.

Y. El Mobariki (✉) · A. Laghrib
ENSA Khouribga, EMI, Université Sultan Moulay Slimane, Slimane, Morocco
e-mail: mobariki2yasmine@gmail.com

A. Laghrib et al. (eds.), *Computational Methods for Inverse Problems and Applications*, Springer Proceedings in Mathematics & Statistics 498,
https://doi.org/10.1007/978-3-031-89498-5_12

Total Variation (TV) regularization and sparsity play complementary roles in image processing and reconstruction, providing important mathematical tools to handle complex visual data [20–22]. TV regularization, defined as the integral of the absolute image gradient throughout the image domain and calculated as the sum of gradients at each pixel, excels at preserving sharp edges while smoothing homogeneous regions. This approach successfully reduces noise without compromising essential image boundaries, making it specially beneficial for piecewise smooth images [23, 24]. Combined with sparsity promotion by L^1-norm regularization, which encourages the solution to focus on the most relevant features while suppressing noise effects, the resulting scheme provides a robust approach to image evaluation. TV manages smooth variations and edge preservation, while sparsity keeps the solution concise and focuses on critical characteristics, ultimately leading to reconstructions that maintain high fidelity while being resilient to noise and artifacts. The mathematical consistency of these two regularization methods has been verified to be mainly powerful in a variety of applications, ranging from medical imaging [23, 25] to computer vision [26], in which the balance between detail maintenance and noise suppression is essential. This combination allows the model to capture both broad, low-frequency information and fine, high-frequency details, making it ideal for tasks requiring high fidelity, such as complex image deconvolution.

Integrating Total Variation and L^1-norm regularization in image processing presents major computational hurdles due to the intrinsic nonsmoothness of both terms, making traditional optimization approaches inadequate for practical applications. When working with large-scale, high-dimensional image data, the complexity increases significantly, needing more advanced algorithmic solutions. The Accelerated Condat-Vũ (ACV) algorithm [1] emerges as a powerful solution to those demanding situations, offering a well constructed framework that efficiently manages the composite character of the optimization problem while retaining theoretical convergence. By leveraging advanced proximal operations and momentum-based acceleration techniques, ACV significantly outperforms conventional methods in terms of convergence speed and computational efficiency, particularly in high-resolution imaging applications. The algorithm's ability to effectively manage structured regularization terms, coupled with its efficient handling of proximal operations, makes it especially valuable in modern image deconvolution tasks where processing speed and solution quality are equally critical, enabling sensible implementation in real-world scenarios where computational resources and time constraints are significant considerations.

This work develops and implements an Accelerated Condat-Vũ approach for image deconvolution, incorporating both Total Variation regularization and sparsity to enhance image quality and clarity. A fundamental contribution of this work is the radical utility of the ACV algorithm in this combined TV and sparse deconvolution process, which provides a tailored solution for high-resolution imaging tasks. Furthermore, this study provides theoretical insights into the convergence characteristics of the ACV algorithm in this context, detailed in Article [1], showcasing enhanced stability and efficiency. Experimental validation in addition supports the effectiveness of the proposed approach, showcasing its resilience to noise and blur in diverse hard scenarios. Together, these contributions boost the field of image deconvolution by handing over a scalable, high-overall performance solution appropriate for programs requiring both detail preservation and noise resilience.

The remainder of this manuscript appears as follows: Sect. 2 describes the model problem and how the ACV technique is applied to our situation. Section 3 delves into the theoretical analysis and rate of convergence. Section 4 outlines the experimental results from this sort of situation, which provide empirical support for the approach's practical utility.

2 Problem Formulation

The proposed model seeks to reconstruct a coherent representation X from noisy and distorted observations h, based on the premise that the image exhibits piecewise smooth characteristics, enforced through Total Variation regularization, and may incorporate sparse features, which is encouraged by the L^1-norm. This formulation is frequently applied in image deconvolution tasks, wherein the objective is to recover an underlying signal image from degraded observations, utilizing both sparsity and smoothness as regularizing factors. In this work, we endeavor to develop an Accelerated Condat-Vũ Algorithm to address the resulting minimization challenge:

$$\min_{X \in \mathcal{X}} \{\|AX - h\|_\alpha + \beta\|DX\|_1 + \mu\|DX\|_\infty\}. \tag{1}$$

In this case, the gradient DX would be the vector of partial derivatives of X with respect to these spatial coordinates. The adjoint operator D^* is often the transpose of the gradient operator D. If D corresponds to a discrete gradient, D^* corresponds to the discrete divergence operator, which sums the differences between adjacent pixel values to compute a measure of change in the opposite direction of the gradient, and the L_∞ norm of the gradient would capture the largest change in X across any spatial direction.

Thus, the gradient DX in a 2D image context would be:

$$DX = \left(\frac{\partial X}{\partial x}, \frac{\partial X}{\partial y}\right)$$

and the infinity norm would be:

$$\|DX\|_\infty = \max\left(\left|\frac{\partial X}{\partial x}\right|, \left|\frac{\partial X}{\partial y}\right|\right).$$

Also:

- $\|AX - h\|_\alpha$: This is the data fidelity term. Here, A is a known linear operator (such as a blurring or convolution operator), X is the unknown image we are trying to reconstruct, and h is the observed (blurred or noisy) image data. The smoothed norm $\|\cdot\|_\alpha$ provides stability and differentiability.
- $\|DX\|_1$: This is the sparsity term, represented by the L^1-norm of the gradient DX, which conserves the discontinuity in edges and blur components in the reconstructed image X. It is often used to promote solutions that are robust to noise or artifacts. The parameter β controls the strength of this sparsity constraint.
- $\mu\|DX\|_\infty$: This is the regularization term, controls the maximum gradient magnitude, providing a way to limit sharp changes in X and helping to manage noise. Here, μ is a regularization parameter that adjusts the effect of this term.

The norm $\|X\|_\alpha$ is defined as:

$$\|X\|_\alpha = \sum_{0 \le i,j \le MN} |X_{i,j}|_\alpha,$$

where $|X|_\alpha$ is defined as a smoothed absolute value function:

$$|X|_\alpha = \frac{1}{\alpha} \log\left(\exp(\alpha X) + \exp(-\alpha X)\right)$$

For further information about this choice, refer to [27]. We define the functions as follows:

$$f_1(X) = \mu\|X\|_\infty + \beta\|X\|_1, \qquad f_2(X) = \|AX - h\|_\alpha.$$

With this formulation, equation (1) becomes:

$$\hat{X} = \underset{X}{\operatorname{argmin}} \quad f_1(DX) + f_2(X) \tag{2}$$

where f_1, f_2 are proper, lower semi-continuous convex functions. Additionally, f_2 is differentiable on $L^2(\Omega)$, where Ω is the image domain with a ε-Lipschitz continuous gradient , D is a linear operator. In the discret setting the image X is represented by a vector belonging to the space $\mathcal{X} = \mathbb{R}^{MN}$, where $M \times N$ represents the image size. To solve the optimization problem defined in (1), we employ the Accelerated Condat-Vũ (ACV) algorithm, described below.

Algorithm 1 Accelerated Condat-Vũ (ACV)

Input: $X_0, W_0 \in \mathcal{X}, Z_0 \in \mathcal{Z}$
Parameters: Step sizes σ_i, γ_i, relaxation parameters λ_i, and acceleration parameters δ_i
for $i = 0, 1, \dots, I - 1$ **do**
  $Y_{i+1} \leftarrow \lambda_i X_i + (1 - \lambda_i) W_i$
  $Z_{i+1} \leftarrow \text{prox}^{\sigma_i}_{f_1^*}\big(Z_i + \sigma_i D(X_i + \delta_i (X_i - X_{i-1}))\big)$
  $X_{i+1} \leftarrow X_i - \gamma_i \nabla f_2(Y_{i+1}) - \gamma_i D^* Z_{i+1}$
  $W_{i+1} \leftarrow \lambda_i X_{i+1} + (1 - \lambda_i) W_i$
end for
$N_I \leftarrow \lambda_I Z_I + (1 - \lambda_I) W_{I-1}$
Output: (W_I, N_I).

Here, the proximal operator is defined as follows:

$$\text{prox}^{\gamma}_{f}(Z) := \text{argmin}_{y \in \mathcal{Z}} \frac{1}{2} \|y - Z\|^2_{\gamma^{-1}} + f(y)$$

where the weighted norm corresponding to the minimizer is expressed as:

$$\|Z\|_{\gamma^{-1}} = \langle \gamma^{-1} Z, Z \rangle.$$

Additionally,

$$f_1^*(y) = \sup_z \langle z, y \rangle - f_1(z),$$

where f_1^* represents the Fenchel conjugate of f_1. Thus, the conjugate of the function f_1 is:

$$f_1^*(y) = \mu \|y\|_1 + \beta \|y\|_\infty$$

The stipulations concerning the parameters further demonstrate the manner in which ACV reinstates the non-accelerated Condat-Vũ algorithm under the condition that $\lambda_i = 1$, and by utilizing the Recommended Parameter Settings delineated in [1], we establish a convergence rate of

$$\mathcal{O}(\frac{\varepsilon}{I^2} + \frac{\|D\|}{I}),$$

as will be elucidated in the subsequent section.

3 Convergence Result of ACV

This section is dedicated to demonstrating the convergence properties of the proposed ACV algorithm. To achieve this, we adapt the methods in [1] to our specific problem and establish the convergence guarantee for ACV using a key lemma.

Let $L : \mathcal{X} \times \mathcal{Z} \to \mathbb{R}$ be the Lagrangian defined as

$$L(X, Z) := \langle DX, Z\rangle - f_1^*(Z) + f_2(X)$$

Additionally, we define the primal-dual gap as

$$G_{\mathbb{B}_1 \times \mathbb{B}_2}(X, Z) := \max_{\hat{Z} \in \mathbb{B}_2} L(X, \hat{Z}) - \min_{\hat{X} \in \mathbb{B}_1} L(\hat{X}, Z)$$

where $\mathbb{B}_1 \times \mathbb{B}_2 \subset \mathcal{X} \times \mathcal{Z}$ is a compact set containing a saddle-point solution.

The subsequent lemma can be directly deduced from ([1], Lemma 3.2):

Lemma 1 *Assume that* $N_{i+1} = \lambda_i Z_{i+1} + (1 - \lambda_i) W_i$.
For each iteration, Algorithm ACV ensures the following inequality holds for all $(X, Z) \in \mathcal{X} \times \mathcal{Z}$*:*

$$\begin{aligned}
&\frac{\sigma_i}{\lambda_i}\left(L(W_{i+1}, Z) - L(X, N_{i+1})\right) \le \frac{\sigma_i(1-\lambda_i)}{\lambda_i}\left(L(W_i, Z) - L(X, N_i)\right)\\
&+ \frac{\sigma_i}{2\gamma_i}(\varepsilon\lambda_i\gamma_i - 1)\|X_{i+1} - X_i\|^2\\
&+ \frac{\delta_i^2\sigma_i^2\|D\|^2}{2}\|X_i - X_{i-1}\|^2 + \frac{\sigma_i}{2\gamma_i}(\|X_i - X\|^2 - \|X_{i+1} - X\|^2)\\
&+ \frac{1}{2}\|Z_i - Z\|^2 - \frac{1}{2}\|Z_{i+1} - Z\|^2 + \sigma_i\delta_i\langle D(X_i - X_{i-1}), Z_i - Z\rangle\\
&- \sigma_i\langle D(X_{i+1} - X_i), Z_{i+1} - Z_i\rangle.
\end{aligned}$$

Finally, we can have the convergence theorem guarantee for ACV:

Theorem 1 *Suppose the algorithm's parameters fulfill the following conditions:*

$$\frac{\sigma_{i+1}(1-\lambda_{i+1})}{\lambda_{i+1}} \le \frac{\sigma_i}{\lambda_i}, \quad \frac{\sigma_{i+1}}{\gamma_{i+1}} \le \frac{\sigma_i}{\gamma_i}, \quad \varepsilon\lambda_i\gamma_i + \sigma_i\gamma_i\|D\|^2 \le 1 \quad \delta_i = \frac{\sigma_{i-1}}{\sigma_i}. \tag{3}$$

The following inequality remains true after $I - 1$ *iterations:*

$$G_{\mathbb{B}_1 \times \mathbb{B}_2}(W_I, N_I) \le \frac{M\lambda_I}{\sigma_I} \tag{4}$$

where

$$M := \max_{(\tilde{X}, \tilde{Z}) \in \mathbb{B}_1 \times \mathbb{B}_2} \frac{\sigma_0(1-\lambda_0)}{\lambda_0}\left(L(W_0, \tilde{Z}) - L(\tilde{X}, N_0)\right) + \frac{\sigma_0}{2\gamma_0}\|X_0 - \tilde{X}\|^2 + \frac{1}{2}\|Z_0 - \tilde{Z}\|^2.$$

Proof According to Lemma 1, we have

$$\frac{\sigma_i}{\lambda_i}(L(W_{i+1},Z)-L(X,N_{i+1}))\le\frac{\sigma_i(1-\lambda_i)}{\lambda_i}(L(W_i,Z)-L(X,N_i))+\frac{\sigma_i}{2\gamma_i}(\varepsilon\lambda_i\gamma_i-1)\|X_{i+1}-X_i\|^2$$
$$+\frac{\delta_i^2\sigma_i^2\|D\|^2}{2}\|X_i-X_{i-1}\|^2+\frac{\sigma_i}{2\gamma_i}(\|X_i-X\|^2-\|X_{i+1}-X\|^2)$$
$$+\frac{1}{2}\|Z_i-Z\|^2-\frac{1}{2}\|Z_{i+1}-Z\|^2+\sigma_i\delta_i\langle D(X_i-X_{i-1}),Z_i-Z\rangle$$
$$-\sigma_i\langle D(X_{i+1}-X_i),Z_{i+1}-Z_i\rangle.$$

By using the conditions in (3) we can find

$$\frac{\sigma_{i+1}(1-\lambda_{i+1})}{\lambda_{i+1}}\big(L(W_{i+1},Z)-L(X,N_{i+1})\big)\le\frac{\sigma_i(1-\lambda_i)}{\lambda_i}(L(W_i,Z)-L(X,N_i))$$
$$+\frac{\sigma_i}{2\gamma_i}(\varepsilon\lambda_i\gamma_i-1)\|X_{i+1}-X_i\|^2$$
$$+\frac{\sigma_{i-1}^2\|D\|^2}{2}\|X_i-X_{i-1}\|^2+\frac{\sigma_i}{2\gamma_i}\|X_i-X\|^2-\frac{\sigma_{i+1}}{2\gamma_{i+1}}\|X_{i+1}-X\|^2$$
$$+\frac{1}{2}\|Z_i-Z\|^2-\frac{1}{2}\|Z_{i+1}-Z\|^2+\sigma_{i-1}\langle D(X_i-X_{i-1}),Z_i-Z\rangle$$
$$-\sigma_i\langle D(X_{i+1}-X_i),Z_{i+1}-Z_i\rangle.$$

Taking the maximum of (X, Z) and summing the last inequality over $i = 0, \ldots, I-1$

$$\frac{\sigma_I}{\lambda_I}G_{\mathbb{B}_1\times\mathbb{B}_2}(W_I,N_I)\le\frac{\sigma_0(1-\lambda_0)}{\lambda_0}\big(L(W_0,\tilde{Z})-L(\tilde{X},N_0)\big)$$
$$+\sum_{i=0}^{I-2}\Big(\frac{\sigma_i}{2\gamma_i}(\varepsilon\lambda_i\gamma_i+\gamma_i\sigma_i\|D\|^2-1)\|X_{i+1}-X_i\|^2\Big)$$
$$+\frac{\sigma_{I-1}}{2\gamma_{I-1}}(\varepsilon\lambda_{I-1}\gamma_{I-1}-1)\|X_I-X_{I-1}\|^2+\frac{\sigma_0}{2\gamma_0}\|X_0-\tilde{X}\|^2-\frac{\sigma_I}{2\gamma_I}\|X_I-\tilde{X}\|^2$$
$$+\frac{1}{2}\|Z_0-\tilde{Z}\|^2-\frac{1}{2}\|Z_I-\tilde{Z}\|^2-\sigma_{I-1}\langle D(X_I-X_{I-1}),Z_I-\tilde{Z}\rangle.$$

We use Young's inequality to establish a bound on the final term:

$$\sigma_{I-1}\langle D(X_I-X_{I-1}),Z_I-\tilde{Z}\rangle\le\frac{\sigma_{I-1}^2\|D\|^2}{2}\|X_I-X_{I-1}\|^2+\frac{1}{2}\|Z_I-\tilde{Z}\|^2$$

it follows that,

$$\frac{\sigma_I}{\lambda_I}G_{\mathbb{B}_1\times\mathbb{B}_2}(W_I,N_I)\le M+\Big(\sum_{i=0}^{I-1}\Big(\frac{\sigma_i}{2\gamma_i}(\varepsilon\lambda_i\gamma_i+\gamma_i\sigma_i\|D\|^2-1)\|X_{i+1}-X_i\|^2\Big)-\frac{\sigma_I}{2\gamma_I}\|X_I-\tilde{X}\|^2$$

where

$$M := \max_{(\tilde{X},\tilde{Z})\in\mathbb{B}_1\times\mathbb{B}_2} \frac{\sigma_0(1-\lambda_0)}{\lambda_0}\Big(L(W_0,\tilde{Z}) - L(\tilde{X},N_0)\Big) + \frac{\sigma_0}{2\gamma_0}\|X_0-\tilde{X}\|^2 + \frac{1}{2}\|Z_0-\tilde{Z}\|^2.$$

By using $\varepsilon\lambda_i\gamma_i + \sigma_i\gamma_i\|D\|^2 \leq 1$ we obtain the desired inequality, completing the proof. □

The convergence inequality $G_{\mathbb{B}_1\times\mathbb{B}_2}(W_I, N_I) \leq \frac{M\lambda_I}{\sigma_I}$ demonstrates that, as the iterations I increase, the algorithm's error term can be controlled by the parameter σ_I. This ensures the stability and efficiency of ACV for large-scale image deconvolution tasks, where both high accuracy and computational efficiency are required.

4 Numerical Part

We outline a series of representative tests, including both simulated and real-world scenarios, to evaluate the performance of our proposed Accelerated Condat-Vũ deblurring technique under various complex degradation operators. These tests are designed to assess the robustness and effectiveness of our method in challenging settings. To streamline computational complexity, we limit the numerical experiments to a single training pair $(\tilde{X}, X_0)$, where $\tilde{X}$ represents the target high-resolution image and X_0 is the 2D interpolation of the initial low-resolution image Y_1, equivalent to $N_t = 1$.

For the numerical setup, we set the lower and upper bounds on α to $\underline{\alpha} = 10^{-3}$ and $\overline{\alpha} = 1$, respectively. Similarly, the values for ω range from $\underline{\omega} = 10^{-1}$ to $\overline{\omega} = 1$. The optimization process employs the BFGS algorithm with the parameters $\varsigma^0 = 0.2$, $\tau = 10^{-6}$, $\rho = 0.14$, $\zeta_1^0 = 0.15$, and $\mathcal{M}_0 = 0.2$. The tolerance for terminating the Algorithm 1 is set to tol $= 10^{-4}$.

In the subsequent sections, we evaluate the denoising performance using two key metrics: the Peak Signal-to-Noise Ratio (PSNR), which is associated with the Mean Squared Error (MSE), and the Structural Similarity Index (SSIM), a metric that better captures perceptual similarity from a human visual perspective. To ensure a fair comparison, all methods are tested using their optimal parameters, selected to achieve the highest PSNR.

In the following numerical simulations, we adopt the relative error E_r as a convergence criterion, defined as:

$$E_r = \frac{\|X^{n+1} - X^n\|_{L^2}}{\|X^n\|_{L^2}} < 10^{-6}, \tag{5}$$

where X^n represents the denoised image at the n-th iteration.

The main algorithm, implemented in MATLAB R2013, generally requires an average of 2 to 3 min to process standard 256×256 grayscale images on an i7 3.4 GHz computer. However, the execution time increases significantly for color images, prompting the exploration of parallel computing methods to improve efficiency.

In the ensuing analysis, we conduct an in-depth comparison of our proposed deblurring approach against several established methods, each renowned for unique strengths in the field. The methods under scrutiny include the Fast gradient-based algorithm (FGBA) approach [24], the Explainable Bilevel Optimization Deblurring (EBOD) method [28], Convex Bregman approach (CBA) [27], and Learning Nonlocal Bilevel approach (LNB) [10]. Notably, parameters for each comparative method were carefully fine-tuned manually, prioritizing optimal visual results. Parameter details for each method are as follows:

- **FGBA method**: We used the Adam optimizer with parameters $\beta_1 = 0.8, \beta_2 = 0.9$, and $\epsilon = 10^{-8}$. The initial learning rate was set to 1×10^{-4}, with a total of 1000 training epochs.
- **EBOD method**: A FISTA-like deblurring algorithm was applied, with α as the vector containing step-length parameters to ensure convergence. The maximum iteration count was fixed at 2000, with a relative error tolerance of $\epsilon = 10^{-4}$.
- **CBA method**: Configured similarly to ADLD, with an adaptive regularization parameter $\epsilon = 0.33$. The contrast parameter β and Gaussian filter size $\sigma \times \sigma$ were adjusted according to the blur type and intensity.
- **LNB algorithm**: Training utilized 3 gradient descent steps, a batch size of 4, and a learning rate of 2×10^{-5}, optimized with the Bilevel Primal Dual algorithm.

This comparative assessment is designed to provide in-depth insights into the strengths and potential limitations of our proposed deblurring approach, relative to leading state-of-the-art techniques. By carefully selecting a diverse set of benchmark images and conducting manual parameter tuning, we ensure a fair and rigorous evaluation, aiming for a comprehensive and meaningful comparison of the resulting visual quality. Each method's performance is scrutinized against specific deblurring challenges, allowing us to derive a detailed understanding of the relative merits and trade-offs of our approach in the context of broader deblurring methodologies.

For the first set of simulated comparison experiments, we utilize three distinct images that vary in size, texture, and complexity. This diversity in image characteristics enables us to assess the robustness and adaptability of the methods under different conditions, providing a more thorough and nuanced evaluation.

- In the first experiment, we used the *Star* sequence and performed two operations on the original image. First, we applied a convolution with a 9×7 Gaussian kernel to introduce blur. Next, we added Gaussian noise with a variance of $\sigma = 25$ and random Poisson noise to the image. The goal was to restore the clean image using the model we proposed. We then compared the results from our model with those obtained using several existing state-of-the-art techniques. The findings are shown in Fig. 1. A careful analysis of the results reveals that our model is competitive with other methods, demonstrating superior image quality, sharper details, and

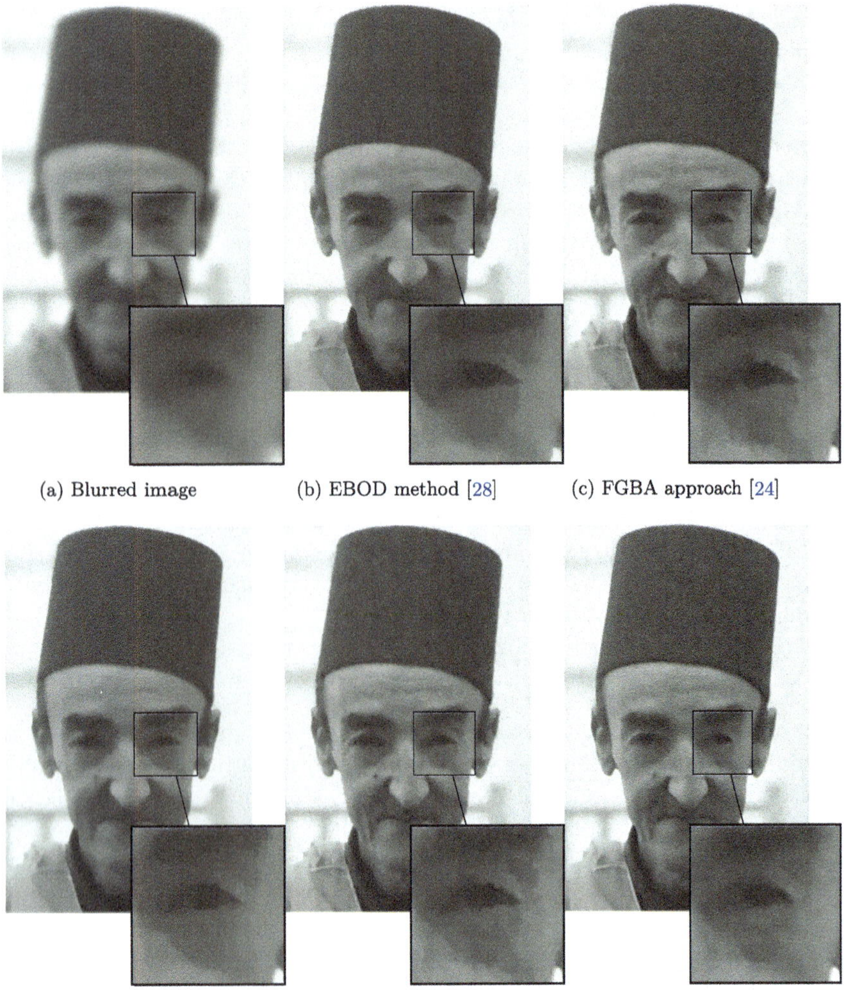

(a) Blurred image (b) EBOD method [28] (c) FGBA approach [24]

(d) CBA [27] (e) LNB [10] (f) Our model

Fig. 1 The deblurring result for the *Man* image using the proposed space variant $w(x)$ and compared to other deblurring approaches

better preservation of important features. While some residual blur remains in our restored image, it is notably suppressed in the restoration produced by the EGNet model.

- *Lion* image: We chose a more intricate and textured image while maintaining the Gaussian blur parameters used in the previous example. However, we increased the challenge by introducing higher noise levels, specifically adding Gaussian noise with a σ value of 30 and a randomize Poisson noise. The outcomes of various deblurring methods applied to recover this image are showcased in Fig. 2.

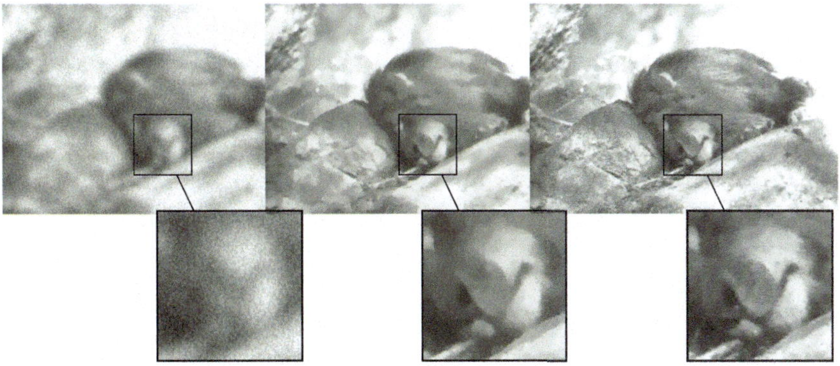

(a) Blurred image (b) EBOD method [28] (c) FGBA approach [24]

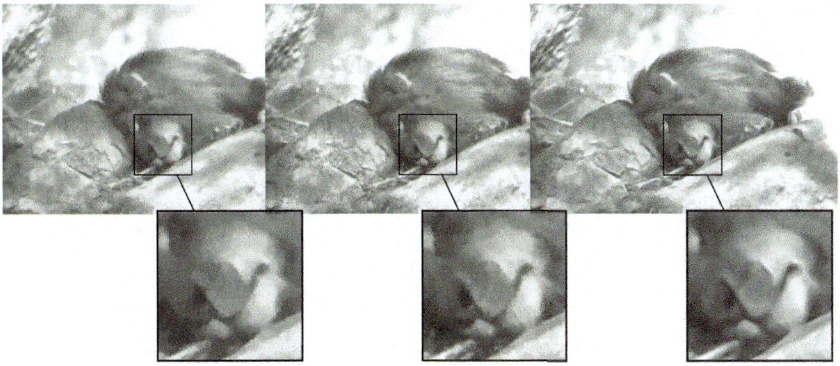

(d) CBA [27] (e) LNB [10] (f) Our model

Fig. 2 The deblurring result for the *Lion* image using the proposed space variant $w(x)$ and compared to other deblurring approaches

Notably, the proposed model stands out in its ability to diminish Gaussian noise compared to alternative approaches. Despite this, both our model and the compared methods exhibit detectable levels of blur in the restored images. In Fig. 3, we plot the evolution of the log-relative error for the tests described above, illustrating how the error decreases over iterations for each method. This plot allows us to visually compare the performance of the various deblurring techniques, highlighting the convergence rates and the effectiveness of the proposed approach relative to existing methods. As shown, our method exhibits a steady reduction in error, demonstrating its efficiency in improving image quality over successive iterations. Notably, the comparison emphasizes the superior performance of our model in terms of error reduction, particularly during the earlier stages of the iterative process.

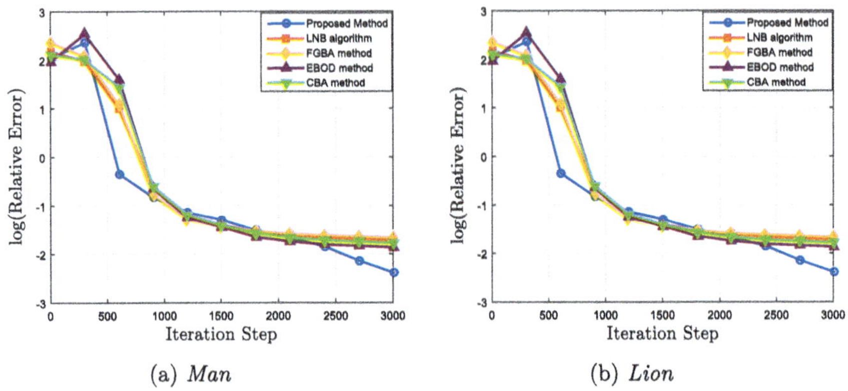

Fig. 3 Evolution of log-relative error across iterations for different deblurring methods

In our quantitative evaluation, we use the PSNR and SSIM metrics to provide objective evidence of the effectiveness and efficiency of our model, demonstrating superior image reconstruction and restoration compared to alternative methods. The corresponding PSNR and SSIM values are compiled in Table 1. Upon analysis, we observe that the inclusion of the proposed regularizer consistently outperforms existing methods in both PSNR and SSIM.

Finally, we evaluate the effectiveness of our regularization model, which is solved by the ACV algorithm, by applying it to the *Book* video sequences. Our approach leverages ACV optimization to accelerate the convergence and reduce the computation of the three minimized functions, thereby improving the model's ability to handle dynamic, time-dependent data inherent in video sequences. The analysis and results presented in Fig. 4 highlight the strengths and advantages of our proposed method in the context of video processing, demonstrating its potential for enhanced performance in handling spatiotemporal variations in video data.

5 Conclusion

We propose an Accelerated Condat-Vũ (ACV) algorithm for deconvolution problems based on Total Variation regularization and sparsity. By formulating the deconvolution problem as a minimization of three convex terms, our model promotes image reconstructions that are both sparse and smooth, effectively suppressing noise and artifacts common in blurred, noisy observations. Our results demonstrate that the

Table 1 The table of PSNR and SSIM values for the above tests and other selected ones. Note that the high value is noted by a bold number in each row

Image	PSNR (SSIM)				
	EBOD method [28]	FGBA approach [24]	CBA [27]	LNB [10]	Proposed approach
Man ($3 \times 3, \sigma = 10$)	33.54 (0.958)	34.49 (0.966)	34.88 (0.960)	**34.97 (0.978)**	34.53 (0.962)
Man ($6 \times 6, \sigma = 30$)	30.44 (0.884)	31.62 (0.899)	32.01 (0.905)	**32.11 (0.912)**	32.08 (0.900)
Man ($10 \times 8, \sigma = 40$)	27.51 (0.796)	27.88 (0.801)	28.22 (0.801)	**28.44 (0.819)**	28.38 (0.810)
Baboon ($5 \times 5, \sigma = 20$)	29.88 (0.881)	29.99 (0.889)	30.37 (0.901)	30.83 (**0.918**)	**30.92** (0.913)
Baboon ($7 \times 6, \sigma = 30$)	28.75 (0.834)	28.82 (0.839)	29.09 (0.855)	**29.81 (0.872)**	29.78 (0.866)
Baboon ($10 \times 10, \sigma = 40$)	25.88 (0.688)	26.43 (0.689)	26.85 (0.693)	27.14 (0.710)	**27.19 (0.722)**
Lion ($8 \times 9, \sigma = 10$)	28.26 (0.785)	29.09 (0.799)	29.56 (0.792)	29.78 (0.808)	**29.89 (0.813)**
Lion ($11 \times 11, \sigma = 20$)	26.55 (0.625)	27.03 (0.669)	27.44 (0.694)	27.97 (0.684)	**28.05 (0.695)**
Lion ($12 \times 9, \sigma = 40$)	24.66 (0.655)	25.04 (0.679)	25.36 (0.792)	25.88 (0.802)	**25.94 (0.810)**
Cameraman ($5 \times 4, \sigma = 20$)	30.52 (0.862)	30.73 (0.874)	31.08 (0.889)	**31.65 (0.893)**	31.63 (0.892)
Cameraman ($9 \times 9, \sigma = 40$)	26.47 (0.652)	26.69 (0.644)	27.18 (0.669)	**27.95 (0.683)**	27.88 (0.678)
Cameraman ($12 \times 12, \sigma = 20$)	24.48 (0.578)	25.72 (0.588)	25.97 (0.594)	26.32 (0.634)	**26.49 (0.651)**
Flower ($4 \times 4, \sigma = 15$)	32.11 (0.921)	32.39 (0.926)	32.74 (0.933)	33.01 (0.937)	**33.17 (0.942)**
Flower ($8 \times 8, \sigma = 25$)	28.54 (0.835)	28.97 (0.845)	29.35 (0.856)	29.76 (0.869)	**29.84 (0.872)**
Flower ($10 \times 10, \sigma = 35$)	25.91 (0.721)	26.14 (0.725)	26.52 (0.734)	26.91 (0.747)	**27.05 (0.752)**
Mountain ($6 \times 6, \sigma = 10$)	31.88 (0.900)	32.10 (0.902)	32.45 (0.911)	32.72 (0.913)	**32.94 (0.915)**
Mountain ($10 \times 9, \sigma = 20$)	29.67 (0.878)	29.95 (0.883)	30.21 (0.889)	30.55 (0.895)	**30.72 (0.898)**
Mountain ($12 \times 10, \sigma = 30$)	27.14 (0.811)	27.46 (0.815)	27.72 (0.820)	27.95 (0.824)	**28.03 (0.826)**
Dog ($6 \times 6, \sigma = 15$)	33.12 (0.892)	33.65 (0.895)	33.92 (0.897)	**34.01 (0.902)**	33.98 (0.900)
Dog ($8 \times 8, \sigma = 25$)	30.34 (0.878)	30.82 (0.883)	31.02 (0.887)	**31.34 (0.890)**	31.27 (0.889)
Dog ($10 \times 10, \sigma = 35$)	27.71 (0.799)	28.14 (0.804)	28.41 (0.811)	28.73 (0.816)	**28.81 (0.818)**
Tiger ($7 \times 7, \sigma = 18$)	32.23 (0.912)	32.51 (0.914)	32.88 (0.920)	33.12 (0.922)	**33.21 (0.923)**
Tiger ($9 \times 9, \sigma = 28$)	29.78 (0.836)	30.02 (0.842)	30.26 (0.847)	30.50 (0.852)	**30.60 (0.855)**
Tiger ($11 \times 11, \sigma = 38$)	27.25 (0.762)	27.50 (0.771)	27.72 (0.775)	27.98 (0.780)	**28.06 (0.784)**

ACV algorithm achieves computational performance and stability, making it suitable for high-dimensional data and real-time applications requiring rapid, accurate image recovery. Future work will explore adaptive parameter choice within the ACV framework, improving robustness across various noise and blur situations.

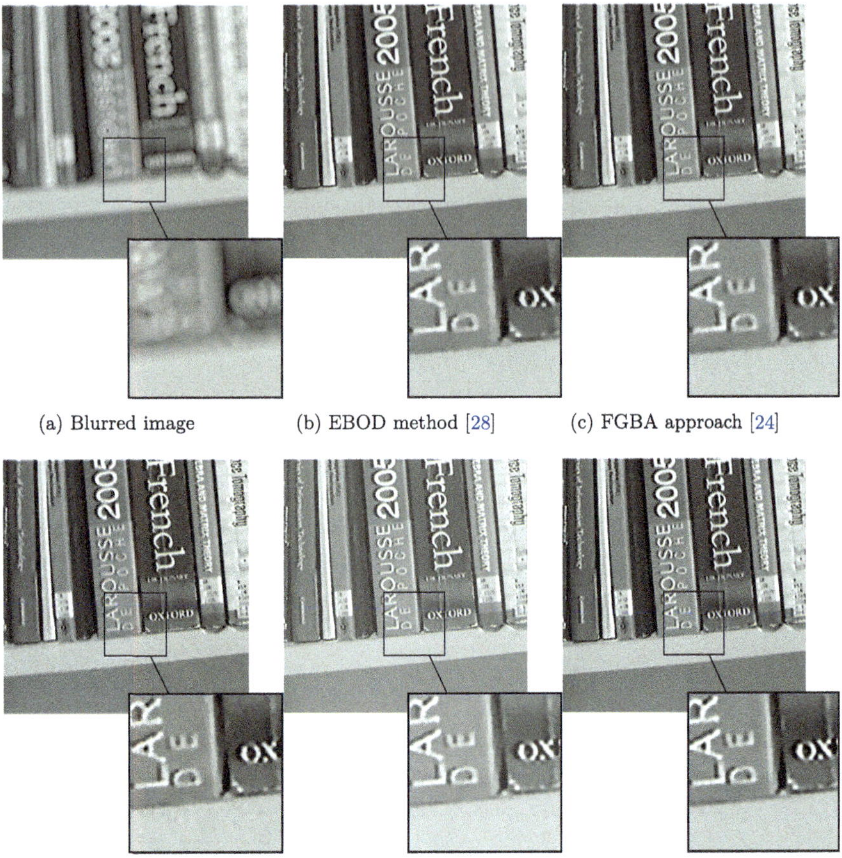

(a) Blurred image (b) EBOD method [28] (c) FGBA approach [24]

(d) CBA [27] (e) LNB [10] (f) Our model

Fig. 4 The deblurring result for the *Books* image using the proposed model and compared to other deblurring approaches

References

1. Driggs, D., Ehrhardt, M.J., Schönlieb, C.-B., Tang, J.: Practical acceleration of the condat-vũ algorithm. SIAM J. Imag. Sci. **17**(4), 2076–2109 (2024)
2. Gregson, J., Heide, F., Hullin, M.B., Rouf, M., Heidrich, W.: Stochastic deconvolution. In: Proceedings of the IEEE Conference on Computer Vision and Pattern Recognition, pp. 1043–1050 (2013)
3. El-Sallam, A.A., Boussaid, F.: Spectral-based blind image restoration method for thin TOMBO imagers. Sensors **8**(9), 6108–6124 (2008)
4. Starck, J.-L., Pantin, E., Murtagh, F.: Deconvolution in astronomy: a review. Publ. Astron. Soc. Pac. **114**(800), 1051 (2002)
5. Boukdir, A., Laghrib, A., Nachaoui, M.: Time-fractionated anisotropic osmosis with bilateral total variation for effective shadow mitigation in image processing. Appl. Math. Model. **137**, 115671 (2025)

6. Nachaoui, A., Laghrib, A., Hadri, A., Nachaoui, M.: A novel variable exponent PDE with dependency on $\gamma(u, |\nabla u0, \sigma|)$ for image despeckling application. Nonlinear Anal. Real World Appl. **83**, 104264 (2025)
7. El Hakoume, A., Zaabouli, Z., Laghrib, A., Afaites, L.: Image restoration using a coupled reaction-diffusion equations. In: New Trends of Mathematical Inverse Problems and Applications, pp. 137–155. Springer, Cham (2023)
8. Laghrib, A., Bella, F.Z.A., Nachaoui, M., Hakim, M.: Learning nonlocal weights for second-order nonlocal super-resolution. Discrete Contin. Dyn. Syst.-S **18**(1), 238–268 (2025). https://doi.org/10.3934/dcdss.2023155
9. Boukdir, A., Nachaoui, M., Laghrib, A.: Hybrid variable exponent model for image denoising: a nonstandard high-order PDE approach with local and nonlocal coupling. J. Math. Anal. Appl. **536**(2), 128245 (2024)
10. El Malki, I., Jauberteau, F., Laghrib, A., Nachaoui, M.: Bilevel learning approach for nonlocal p-Laplacien image deblurring with variable weights parameter w (x). J. Vis. Commun. Image Represent. **103**, 104248 (2024)
11. El Malki, I., Jauberteau, F., Laghrib, A., Nachaoui, M.: A bilevel learning approach for nonlocal image deblurring with variable weights parameter. Digit. Signal Process. **149**, 104505 (2024)
12. El Malki, I., Jauberteau, F., Laghrib, A., Nachaoui, M.: Enhancing deblurring through optimal controlled bilevel with nonlocal weighted model. Math. Control Relat. Fields (2024)
13. Nachaoui, M., Laghrib, A., El Hakoume, A.: On theoretical result of a controlled $p(x)$-reaction diffusion equation for Poisson noise reduction. Evol. Equ. Control Theory **13**(1), 280–304 (2024)
14. Nachaoui, M., Laghrib, A., Hakim, M.: A new space-variant optimization approach for image segmentation. In: International Conference on Numerical Analysis and Optimization Days, pp. 87–97. Springer (2021)
15. Alahyane, M., Hakim, A., Laghrib, A., Raghay, S.: Fluid image registration using a finite volume scheme of the incompressible Navier stokes equation. Inverse Probl. Imaging **12**(5) (2018)
16. Alahyane, M., Hakim, A., Laghrib, A., Raghay, S.: An improved simplec scheme for fluid registration. Math. Model. Anal. **28**(1), 71–90 (2023)
17. Laghrib, A., Ghazdali, A., Hakim, A., Raghay, S.: A multi-frame super-resolution using diffusion registration and a nonlocal variational image restoration. Comput. Math. Appl. **72**(9), 2535–2548 (2016)
18. Afraites, L., Hadri, A., Laghrib, A., Nachaoui, M.: A weighted parameter identification PDE-constrained optimization for inverse image denoising problem. Vis. Comput. **38**(8), 2883–2898 (2022)
19. Laghrib, A., Hadri, A., Hakim, A.: An edge preserving high-order PDE for multiframe image super-resolution. J. Franklin Inst. **356**(11), 5834–5857 (2019)
20. Laghrib, A., Afraites, L.: Image denoising based on a variable spatially exponent PDE. Appl. Comput. Harmon. Anal. **68**, 101608 (2024)
21. Laghrib, A., Afraites, L., Nachaoui, M.: New Trends of Mathematical Inverse Problems and Applications: ICNTAM 2022, Béni Mellal, Morocco, 19–21 May 2023, vol. 428. Springer (2023)
22. Laghrib, A., Afraites, L., Nachaoui, M.: New trends of mathematical inverse problems and applications (2023)
23. Xie, H., Lv, Y.: Performance evaluation of a smoothed total variation regularization algorithm. In: IEEE Nuclear Science Symposium, Medical Imaging Conference and International Symposium on Room-Temperature Semiconductor Detectors (NSS MIC RTSD), pp. 1–1. IEEE (2023)
24. Beck, A., Teboulle, M.: Fast gradient-based algorithms for constrained total variation image denoising and deblurring problems. IEEE Trans. Image Process. **18**(11), 2419–2434 (2009)
25. Wang, T., He, M., Shen, K., Liu, W., Tian, C.: Learned regularization for image reconstruction in sparse-view photoacoustic tomography. Biomed. Opt. Express **13**(11), 5721–5737 (2022)

26. Estrela, V.V., Magalhaes, H.A., Saotome, O.: Total variation applications in computer vision. In: Handbook of Research on Emerging Perspectives in Intelligent Pattern Recognition, Analysis, and Image Processing, pp. 41–64. IGI Global (2016)
27. Unger, M., Pock, T., Werlberger, M., Bischof, H.: A convex approach for variational super-resolution. In: Pattern Recognition: 32nd DAGM Symposium, Darmstadt, Germany, 22–24 Sept 2010. Proceedings 32, pp. 313–322. Springer (2010)
28. Bonettini, S., Franchini, G., Pezzi, D., Prato, M.: Explainable bilevel optimization: an application to the Helsinki deblur challenge. Inverse Probl. Imaging **17**(5), 925–950 (2023)

A Deep Numerical Study of BFGS and LBFGS Methods for Solving Optimization Problems Arising from Inverse Applications

Mohamed Ziouane, Hafida Hamdi, Mourad Nachaoui, and Abdeljalil Nachaoui

Abstract Optimization algorithms are designed to find optimal solutions by minimizing or maximizing an objective function subject to constraints. The objective function, which may be non-linear, complex, or non-differentiable, defines the relationship between the system parameters and the desired outcome. This work focuses on numerical optimization techniques, specifically comparing the efficiency of Gradient Descent (GD), Broyden-Fletcher-Goldfarb-Shanno (BFGS), and Limited-memory BFGS (LBFGS) algorithms for problems arising from inverse or ill-conditioned scenarios.

Keywords Optimization · LBFGS · BFGS · GD · Problem inverse

1 Introduction

Optimization problems are present in many fields, from engineering and economics to data science and artificial intelligence [1–4]. Basically, an optimization problem consists in finding the best solution among a set of possible solutions, by minimizing or maximizing an objective function, while respecting certain constraints. A minimization problem can be formulated as follows

$$\min_{x \in \mathcal{X}} f(x), \tag{1}$$

Hafida Hamdi, Mourad Nachaoui, and Abdeljalil Nachaoui are contributed equally to this work.

M. Ziouane (✉) · H. Hamdi · M. Nachaoui
Beni-Mellal, Morocco
e-mail: ziouane.med1@gmail.com

M. Nachaoui
e-mail: m.nachaoui@usms.ma

A. Nachaoui
Nantes, France
e-mail: abdeljalil.nachaoui@univ-nantes.fr

A. Laghrib et al. (eds.), *Computational Methods for Inverse Problems and Applications*, Springer Proceedings in Mathematics & Statistics 498,
https://doi.org/10.1007/978-3-031-89498-5_13

where f is the objective function, that we wish to minimize and $\mathcal{X}$ generally represents the set of possible solutions x. Minimization problems can be classified into several categories, including: nonlinear problems, where one or the other (or both) of the objective function or constraints are nonlinear, convex problems, here any local minimum region is also a global minimum and non-convex problems, where several local minima can exist. The important features for all optimization algorithms are as follows [5]:

- Robustness: optimization methods should perform well for a wide variety of problems in their class and for all sensible values of initial estimates.
- Efficiency: optimization methods should not require excessive computer storage or time.
- Accuracy: optimization methods should produce accurate solutions. These features can conflict. For example, a robust method may be the slowest, while a method that achieves rapid convergence over iterations may require excessive computer storage space. The trade-offs between the aforementioned optimization features constitute one of the main challenges of numerical optimization.

Several inverse problems are solved as unconstrained optimization problems. They aim to deduce unknown parameters or functions from observed data. While many inverse problems involve constraints [6–13], some can be effectively approached using unconstrained methods [14].

The data fitting term has a ill-conditioned Hessian after discretization, which also produces a dense Hessian. In fact, even matrix-vector multiplications are computationally expensive, whereas the Hessian of the regularizer is positive definite, well conditioned, and sparse, with computationally inexpensive matrix-vector products.

To solve this type of inverse problem, there are two categories:

- Global optimization, such as evolutionary algorithms [15–18] and their variants.
- Local optimization, which is divided into two subcategories:
  - First-order optimization, such as gradient descent and its variants.
  - Second-order optimization, such as the Newton method and its variants.

As we are interested in second-order optimization, the BFGS method [19] and its variants emerge as the best methods to use for optimization problems.

In this work, our objective is to evaluate the robustness and performance of LBFGS [20] for solving optimization problems arising from inverse problems. The main objective is to study the behavior of the data-fitting term in the optimization problem.

2 Gradient Descent

Gradient descent [21] is a method for minimizing an objective function $f(x)$ parameterized by the parameters of a model $x \in \mathcal{X}$ by updating the parameters in the direction opposite to the gradient of the objective function $\nabla f(x)$ with respect to the parameters. The learning rate α determines the size of the steps we take to reach a (local) minimum. In other words, we follow the direction of the slope of the surface created by the objective function downward until we reach a valley.

2.1 Gradient Descent Variants

Gradient descent algorithms can be categorized into three variants based on the quantity of data used to calculate the objective function's gradient. This choice involves a trade-off between the precision of the parameter update and the computational cost associated with each update.

2.1.1 Batch Gradient Descent

Batch gradient descent calculates the gradient of the cost function with respect to the parameters x using the entire training dataset:

$$x_{k+1} = x_k - \alpha \nabla f(x_k), \tag{2}$$

Since batch gradient descent requires the calculation of gradients over the complete dataset for each parameter update, it can be prohibitively slow, especially for large datasets that do not fit in memory.

For a predefined number of epochs, we first calculate the gradient of the loss function for the entire dataset with respect to the parameters. If you derive the gradients yourself, it is recommended to verify them. We then update our parameters in the direction of the gradients, with the learning rate alpha determining the magnitude of the update we perform. Batch gradient descent is guaranteed to converge to the global minimum for convex error surfaces and to a local minimum for non-convex surfaces.

2.1.2 Stochastic Gradient Descent (SGD)

Stochastic Gradient Descent [22] is a widely used optimization algorithm in machine learning, particularly for training neural networks. Unlike batch gradient descent, which calculates the gradient using the entire dataset, SGD computes the gradient using a random subset of the dataset, known as a mini-batch. SGD can be susceptible to noise in the data, potentially leading to unstable convergence. Additionally, selecting an appropriate learning rate is critical for SGD's performance. Moreover, SGD may become trapped in local minima, particularly for non-convex optimization problems.

2.1.3 Mini-batch Gradient Descent

Mini-batch gradient descent [23] is an optimization algorithm used in machine learning to update model parameters iteratively. It calculates the error and updates weights using small batches of data, offering a balance between speed and stability compared to batch and stochastic gradient descent.

While mini-batch gradient descent is an efficient optimization algorithm, it comes with several challenges. Firstly, selecting an appropriate learning rate is crucial. A learning rate that is too small results in painfully slow convergence, whereas a learning rate that is too large can hinder convergence or even cause the loss function to diverge. Secondly, learning rate schedules, which dynamically adjust the learning rate during training, often have limitations. These schedules are typically predefined and may not adapt well to the specific characteristics of a given dataset. Moreover, this optimization algorithm is susceptible to becoming trapped in local minima or saddle points, which can significantly slow down the convergence process.

2.2 *Gradient Descent Algorithm*

Given an initial iterate $x_0 \in \mathcal{X}$ and a learning rate $\alpha > 0$, the gradient algorithm defines a sequence of iterates $(x_0, \ldots, x_m) \in \mathcal{X}$ until a stopping criterion is met. It moves from x_k to x_{k+1} through the following steps (Fig. 1).

GD is a first-order iterative method to find a local minimum of a differentiable function (the cost function). It was first proposed by Cauchy in 1847 and further developed by Hadamard in 1907. The following theorem proves the convergence of the GD algorithm to a local minimum x^* of f in $\mathcal{X}$.

Theorem 2.1 (GD Algorithm) *Let $f : \mathcal{X} \to \mathbb{R}$ be a differentiable function. Then, there exists a local minimum that can be reached by iterative updates according to the following system:*

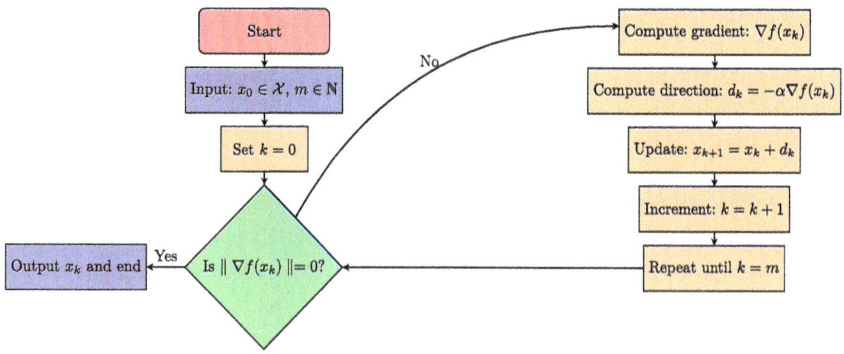

Fig. 1 Algorithm GD

$$\begin{cases} x_0 = x \\ x_{k+1} = x_i - \alpha \nabla f(x_k) \end{cases} \tag{3}$$

where α is called the learning rate.

Proof The goal is to update the values of a given function f, where $x = (x_1, \ldots, x_m)$, to reach a local minimum. GD employs an iterative descent approach, moving in the direction of the negative gradient $-\nabla f(x)$ to minimize $f(x)$ as rapidly as possible $\nabla f(x) = \left(\frac{\partial f(x)}{\partial x_1}, \ldots, \frac{\partial f(x)}{\partial x_m}\right) = \left(f'(x_1), \ldots, f'(x_m)\right) \approx$
$\approx \left(\lim_{h\to 0} \frac{f(x_1+h)-f(x_1)}{h}, \ldots, \lim_{h\to 0} \frac{f(x_m+h)-f(x_m)}{h}\right).$

Ignoring the coordinate-wise description, therefore $\nabla f(x) = \lim_{h\to 0} \frac{f(x+h)-f(x)}{h}$
$\Rightarrow \nabla f(x) \approx \dfrac{f(x+h)-f(x)}{h} \Rightarrow f(x+h) \approx f(x) + h\nabla f(x).$

For a small positive step size α, let $h = -\alpha \nabla f(x)$ Since $\alpha > 0$ and $(\nabla f(x))^2 \geq 0$ we have: $f(x - \alpha \nabla f(x)) \approx f(x) - \alpha \nabla f(x) \nabla f(x) = f(x) - \alpha (\nabla f(x))^2 \leq f(x)$.

In other words, the function f takes on smaller values $f(x - \alpha \nabla f(x)) \leq f(x)$ for inputs $x - \alpha \nabla f(x)$ as long as α remains positive. Recall that, as before, these expressions should be considered component-wise.

The expression $f(x - \alpha \nabla f(x)) \leq f(x)$, can be equivalently expressed as: $f(\underbrace{x - \alpha \nabla f(x)}_{x^1}) \leq f(\underbrace{x}_{x^0})$. Consequently, the sequence that updates the minimal values of the function f with initial value x_0 and positive scaling factors α is given by: $x_{k+1} = x_k - \alpha \nabla f(x_k)$. □

3 Broyden-Fletcher-Goldfarb-Shanno Method (BFGS)

The method commonly used to solve problem (1) is the BFGS [19, 24] method, which serves as the inspiration for several recent optimization algorithms. The BFGS update possesses self-correcting properties [25], which makes it the most effective compared to other methods in terms of error precision and convergence speed.

The updates will use estimates of the gradient $\nabla f(x_k)$. The update rule for the algorithm will take the form

$$x_{k+1} = x_k - \alpha_k \Lambda_k \nabla f(x_k) \qquad \text{for} \quad k \in \{1, \ldots, m\},$$

Newton's method uses Λ_k as the inverse Hessian $(\nabla^2 f(x_k))^{-1}$, while in the gradient method, it is the identity matrix. For BFGS [24, 26], Λ_k will be an approximation of the inverse Hessian. Many parameters are needed for this algorithm, such as: a step size α_k ((6) and (7)), a positive integers m, a positives reals numbers (η, σ) and initialize iterative $x_0 \in \mathcal{X}$. For each iteration, the algorithm computes the gradient and updates the inverse Hessian approximation Λ_k. The vector s_k records the average direction in which the algorithm has made progress over the past m iterations $s_k = \alpha_k d_k$. The vector y_k is obtained by taking the difference between successive gradients $y_k = \nabla f(x_{k+1}) - \nabla f(x_k)$. The inverse Hessian approximation Λ_k is defined from the pairs (s_k, y_k) for $k \in \{1, \ldots, m\}$.

3.1 The Inverse Hessian Approximation

The operator Λ_k is obtained as $\Lambda_k = \Lambda_k^{(m)}$ from the seed matrix $\Lambda_k^{(0)}$ and the storage $(s_j, y_j)_{j=0}^{j=m}$ from the recursion:

$$\Lambda_k^{(j+1)} = (I - \frac{y_j s_j^T}{y_j^T s_j})^T \Lambda_k^{(j)} (I - \frac{y_j s_j^T}{y_j^T s_j}) + \frac{s_j s_j^T}{y_j^T s_j}, \quad j = 0, \ldots, m \tag{4}$$

Note that the update in Eq. (4) preserves positive definiteness ($y_j^T s_j > 0$), which implies that Λ_k and each $\Lambda_k^{(j)}$ will be positive definite, as will their inverses.

From the Sherman-Morrison [27] formula we infer that H_k can be obtained by setting $H_k = H_k^{(m)}$ in the recursion

$$H_k^{(j+1)} = H_k^{(j)} - \frac{H_k^{(j)} s_j (H_k^{(j)} s_j)^T}{(H_k^{(j)} s_j)^T s_j} + \frac{y_j y_j^T}{y_j^T s_j}, \quad \textbf{for } j = 0, 1, \ldots, m, \tag{5}$$

where

$$s_k = \alpha_k d_k \quad \text{and} \quad y_k = \nabla f(x_{k+1}) - \nabla f(x_k)$$

and: $H_k^{(0)} = (\Lambda_k^{(0)})^{-1}$.

3.2 The Wolfe-Powell and Armijo Conditions

A line search that ensures the satisfaction of the Wolfe-Powell [26] conditions or the strong Wolfe-Powell conditions is necessary for updating the BFGS algorithm. That said, the existence of constants $\sigma \in (0, 1)$ and $\eta \in [\sigma, 1)$ such that the step sizes α_k satisfy the Armijo [28] condition is required.

$$f(x_{k+1}) \leqslant f(x_k) + \alpha_k \sigma \nabla f(x_k)^T d_k \tag{6}$$

and either the curvature condition or the strong curvature condition:

$$\nabla f(x_{k+1})^T d_k \geqslant \eta \nabla f(x_k)^T d_k \text{ respectively } |\nabla f(x_{k+1})^T d_k| \geqslant \eta |\nabla f(x_k)^T d_k| \tag{7}$$

3.3 Algorithm of BFGS Method

3.4 Convergence Analysis

Now we are ready to state the main convergence result for the BFGS algorithm, under the same assumptions as presented in [29] (Fig. 2).

Theorem 3.1 ([29], Theorem 3.2) *Suppose that f is bounded below in $\mathcal{X}$, f is continuously twice differentiable in an open set $\mathcal{M}$ containing the level set $\mathcal{L} = \{x \in \mathcal{X} : f(x) \leq f(x_0)\}$. Then the sequence generated by Algorithm (2) is globally convergent in the sense that*

$$\lim_{k\to\infty} \|\nabla f(x_k)\| = 0. \tag{8}$$

In addition, if

$$\|s_k\| = o\left(\frac{1}{k^{1+\varepsilon}}\right) \quad \text{for any} \quad \varepsilon > 0,$$

then the sequence generated by Algorithm (2) converges to some local minimum point x^* satisfying $\|\nabla f(x^*)\| = 0$ with superlinear rate.

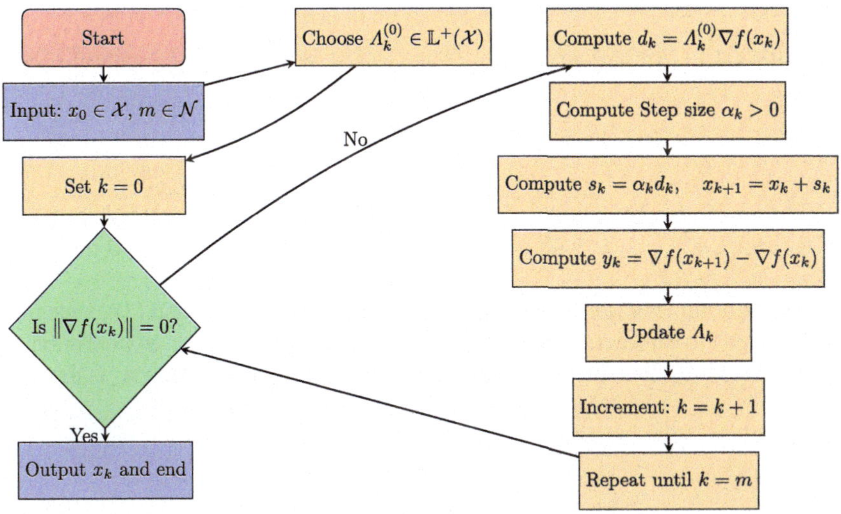

Fig. 2 BFGS algorithm

4 Limited-Memory BFGS (LBFGS)

Given the widespread use of the standard BFGS method, most studies on limited-memory methods focus on the Limited-memory BFGS (LBFGS) algorithm [30, 31]. It is a variant of BFGS designed for large-scale optimization problems where the $O(d^2)$ cost of storing and updating the inverse Hessian matrix Λ would be prohibitively expensive. In LBFGS, the estimation of the inverse Hessian is based solely on the last M steps in the parameter and gradient space, the quasi-Newton direction is obtained directly from these via a matrix-free approach. This reduces the computational cost in both space and time per iteration.

Unlike gradient descent, LBFGS does not immediately lend itself to a stochastic version. The updates in the stochastic gradient method average together to produce a downhill direction in expectation. However, as pointed out in Byrd et al. [32], the updates used in LBFGS to construct the inverse Hessian approximation overwrite one another instead of averaging.

In this part, we will present a globalization of the LBFGS method [33]. This is the first method that converges globally for nonconvex objective functions while also recovering the original LBFGS method under local strong convexity.

The LBFGS update shares the same structure as BFGS. The primary difference lies in the initial update of $\Lambda_k^{(0)}$ and the memory size M. The two most commonly used scaling factors for $\Lambda_k^{(0)} = \gamma_k I$ in LBFGS methods are the BB scaling factors introduced by Barzilai and Borwein [34]. The γ_k is given by the Definition 4.1:

Definition 4.1 (*BB scaling*) Let $k \in \mathbb{N}$ and let $(s_k, y_k) \in \mathcal{U}_{ad} \times \mathcal{U}_{ad}$ with $y_k^T s_k > 0$. We set

$$\gamma_{k+1}^{-} := \frac{y_k^T s_k}{\| y_k \|^2} \textit{ and } \gamma_{k+1}^{+} := \frac{\| s_k \|^2}{y_k^T s_k}$$

To simplify the presentation of the LBFGS method, we restrict attention to choices of the form $\Lambda_k^{(0)} = \gamma_k I$, with $\gamma_k \in \left[\gamma_k^{-}, \gamma_k^{+}\right]$. For memory size $M = 0$ this yields $d_k = -\gamma_k \nabla f(x_k)$.

The algorithm requires several parameters: a step size α_k ((6) and (7)), a memory size M, a positive integers m, a positives reals numbers $(c_0, c_1, c_2, \eta, \sigma)$, and initialize iterative $x_0 \in \mathcal{X}$. Construction of the Inverse Hessian Approximation Λ_k is the same as (5).

4.1 Algorithm of LBFGS Method

See Fig. 3.

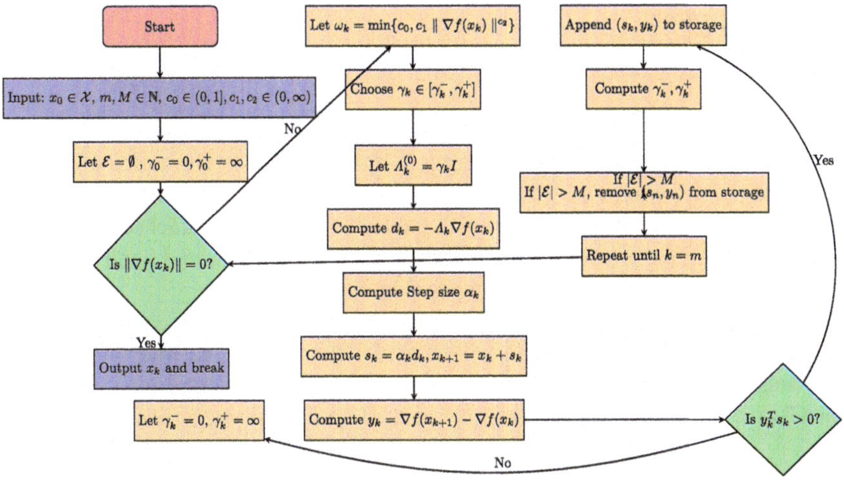

Fig. 3 Algorithm LBFGS

4.2 *Convergence Analysis*

The convergence guarantees of the new method and the transition to the classical LBFGS method rely on a new form and a careful calibration of the cautious updates introduced for BFGS by Li and Fukushima in [35].

Theorem 4.10 in [33] demonstrates that, under certain assumptions, the Wolfe-Powell and Armijo conditions guarantee global convergence, making the underlying analysis particularly interesting.

Theorem 4.1 ([33], Theorem 4.10) *The Algorithm LBFGS either terminates after finitely many iterations with an x_k that satisfies $\nabla f(x_k) = 0$ or it generates a sequence (x_k) with*

$$\lim_{k\to\infty} \|\nabla f(x_k)\| = 0 \tag{9}$$

In particular, every cluster point of (x_k) is stationary.

5 Numerical Results

A comparative study will be conducted to evaluate the performance of gradient descent, BFGS, and LBFGS methods in order to identify the most effective optimization algorithm for optimization problems. Performance metrics will include accuracy, error calculation, and robustness to noise, as assessed on a variety of matrix types and sizes. The findings of this study will enable us to select the most efficient method for a specific application.

5.1 Example 1

We consider the matrix $A = \begin{pmatrix} 3 & 2 \\ 1 & 2 \end{pmatrix}$, whose condition number is acceptable, $cond(A) = 4.26$, and the vector b defined by: $b^t = \left(5, 3\right)$. The exact solution to the system $Ax = b$ is: $x^t = \left(1, 1\right)$.

Just after 100 iterations, the three methods converge to the exact solution of the system, with better precision for BFGS, followed by LBFGS, and finally GD. The following table summarizes these results (Fig. 4).

The BFGS and LBFGS methods are generally more effective at converging to the solution, especially in the case of small matrices. They use second-order information, allowing them to adjust their search direction for the minimum more quickly, whereas the GD method requires a large number of iterations to achieve a satisfactory result.

5.2 Example 2

We consider the matrix $A = \begin{pmatrix} 10 & 7 & 8 & 7 \\ 7 & 5 & 6 & 5 \\ 8 & 6 & 10 & 9 \\ 7 & 5 & 9 & 10 \end{pmatrix}$, whose condition number is acceptable, $cond(A) = 2984.09$, and the vector b defined by: $b^t = \left(32, 23, 33, 31\right)$. The exact solution to the system $Ax = b$ is: $x^t = \left(1, 1, 1, 1\right)$.

After 100 iterations, with and without noise of (± 0.1), we observe that both methods remain robust in convergence, while GD explodes towards infinity. This demonstrates the weakness of GD in problems with very high conditioning (Fig. 5).

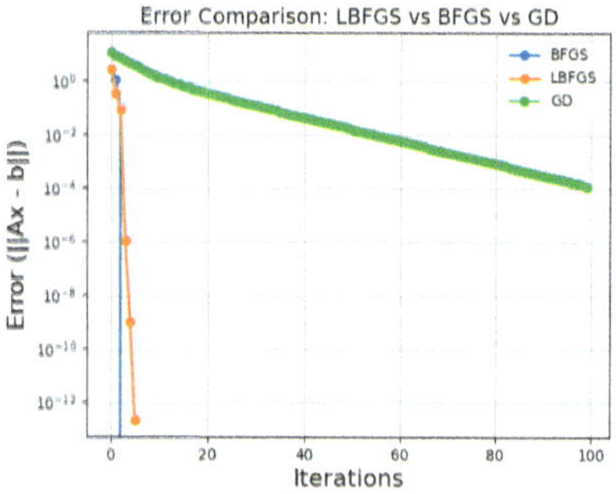

Error of	GD	BFGS	LBFGS
100 Iterations	$2.53\,e^{-4}$	0.0	0.0

Fig. 4 Error comparison: LBFGS versus BFGS versus GD

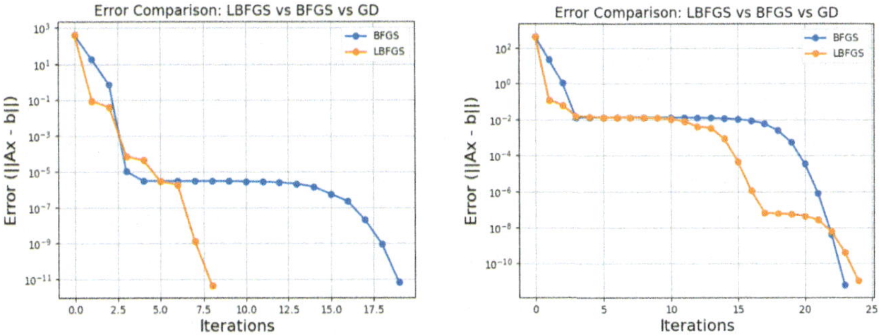

(a) 100 iterations (b) 100 iterations with noise of ±0.1

Fig. 5 Error comparison with and without noise: LBFGS versus BFGS versus GD

Iterations/Error	GD	BFGS	LBFGS
100 iterations	$1.052\,e^{84}$	$1.42\,e^{-13}$	$9.52\,e^{-12}$
100 iterations with noise ±0.1	$1.052\,e^{84}$	$1.30\,e^{-13}$	$2.33\,e^{-13}$

5.3 *Example 3: Hilbert's Matrix*

We consider Hilbert's matrix: $H_n = \left[\frac{1}{i+j-1}\right]_{1\le i,j\le n}$ for $n \in \{10, 20\}$, which has a large condition number, and the vector b defined by: $b = (1)_{i=1}^{n}$.

Three tests will be performed to assess the efficiency of each method in terms of the number of iterations and convergence error. The convergence errors obtained after 100 and 10,000 iterations using GD, BFGS, and LBFGS methods, respectively, with matrices of different sizes (i.e., H_{10} , the Hilbert matrix of size 10 and H_{20}, the Hilbert matrix of size 20) are presented in the following table:

Iterations/Error	GD	BFGS	LBFGS	Condition number
H_{10} with 100 iterations	1.53	$6.59\,e^{-4}$	$6.59\,e^{-4}$	$1.602\,e^{+13}$
H_{20} with 100 iterations	2.24	$6.21\,e^{-4}$	$6.21\,e^{-4}$	$1.148\,e^{+18}$
H_{20} with 1000 iterations/noise= ±0.1	1.16	$9.02\,e^{-2}$	$9.02\,e^{-2}$	$1.148\,e^{+18}$

Although Hilbert matrices are ill-conditioned, the BFGS and LBFGS methods demonstrates its robustness in terms of convergence, even for different numbers of iterations, when compared to the gradient descent and BFGS methods (Fig. 6).

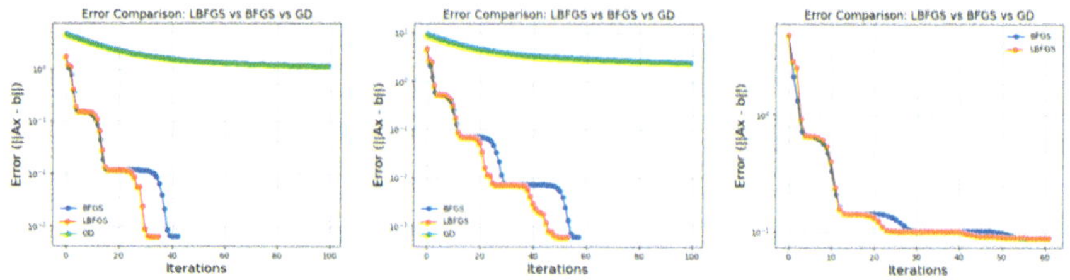

(a) H_{10} (b) H_{20} (c) H_{20}, noise ± 0.1

Fig. 6 Error comparison: LBFGS versus BFGS versus GD

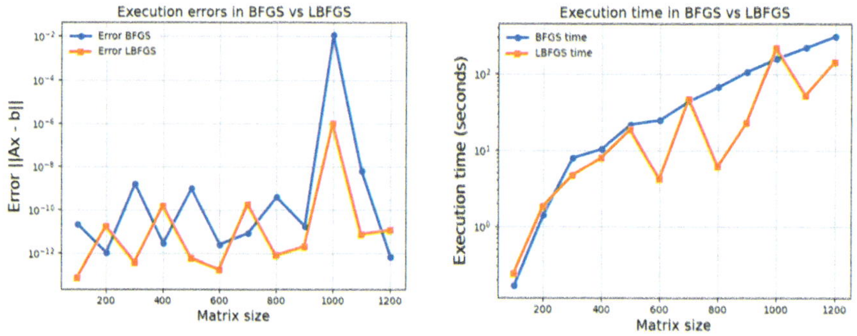

Fig. 7 Error and time comparison: LBFGS versus BFGS

5.4 *Example 4*

For this example, we will use random matrices of size $n \times n$ of the form $A = np.random.randn(n, n)$ and the vector $b = np.ones(n)$ with sizes $n \in \{100k\}$, where $k \in \{1, 2, \ldots, 15\}$. The following figures present the results of this experiment (Fig. 7).

For a set of full, random matrices with sizes ranging from 100 to 1500, which makes the minimum search more challenging, both methods LBFGS and BFGS demonstrated robustness in terms of convergence. In contrast, the GS method failed to converge and rapidly diverged, even for large and ill-conditioned matrices. Among the methods compared, LBFGS proved to be the most dominant, converging more quickly than BFGS. Therefore, we conclude that the LBFGS method is generally superior to BFGS, especially for large-scale problems.

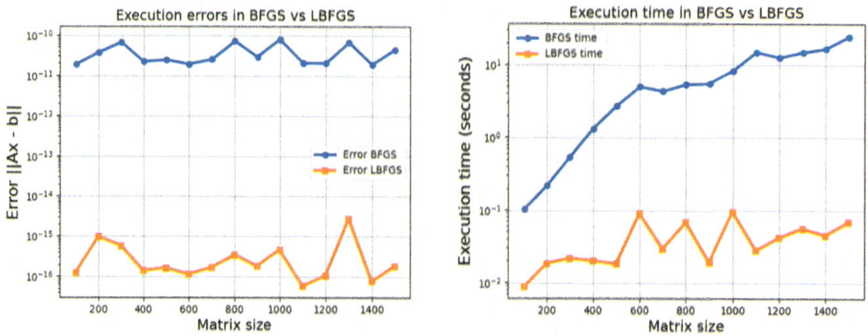

Fig. 8 Error and time comparison: LBFGS versus BFGS

5.5 *Example 5: Tridiagonal Matrix*

For this example, we will use a Tridiagonal matrices of the form:

$$A = \begin{bmatrix} 4+\varepsilon_1 & -1 & 0 & . & 0 \\ -1 & 4+\varepsilon_2 & -1 & . & . \\ 0 & . & . & . & 0 \\ . & . & -1 & 4+\varepsilon_{n-1} & -1 \\ 0 & . & 0 & -1 & 4+\varepsilon_n \end{bmatrix}$$

where $-1 < \varepsilon_i < 1$ has a random value, and the vector $b = np.ones(n)$ with sizes $n \in \{100k\}$, where $k \in \{1, 2, \ldots, 15\}$. The following figures present the results of this experiment (Fig. 8).

The results show interesting trends in the comparison of the efficiency of the BFGS and LBFGS methods in terms of convergence and computation time for different problem sizes n. The LBFGS achieves much smaller error values than BFGS in all cases, often on the order of 10^{-16}, which is very close to zero. In contrast, BFGS generally exhibits slightly larger error values than LBFGS, on the order of 10^{-11}. In terms of convergence time, LBFGS is significantly faster than BFGS in all cases. As n increases, the gap in execution time becomes more pronounced. For example, for $n = 1000$, BFGS takes approximately 8.35 s, while LBFGS only takes 0.0974 s. In summary BFGS is more expensive in terms of memory and computation time as the problem size increases. Therefore, it is better suited for moderate-sized problems or situations where memory is less constrained, but it becomes less efficient for large-dimensional problems.

6 Conclusion

Numerical experiments demonstrate the superior performance of the LBFGS method compared to GD and BFGS, especially when handling ill-conditioned matrices, such as Hilbert matrices, tridiagonal matrices, and large-sized matrices. LBFGS offers a favorable balance between convergence speed and accuracy, often achieving satisfactory solutions in fewer iterations. Moreover, LBFGS exhibits remarkable robustness to the ill-conditioned nature of Hilbert matrices, maintaining stable convergence even in the presence of increased noise. In conclusion, LBFGS emerges as a reliable and efficient optimization technique, particularly well-suited for challenging problems involving ill-conditioned matrices.

References

1. El Mourabit, I., El Rhabi, M., Hakim, A., Laghrib, A., Moreau, E.: A new denoising model for multi-frame super-resolution image reconstruction. Signal Process. **132**, 51–65 (2017)
2. Laghrib, A., Ghazdali, A., Hakim, A., Raghay, S.: A multi-frame super-resolution using diffusion registration and a nonlocal variational image restoration. Comput. Math. Appl. **72**(9), 2535–2548 (2016)
3. Nachaoui, M., Laghrib, A.: An improved bilevel optimization approach for image super-resolution based on a fractional diffusion tensor. J. Franklin Inst. **359**(13), 7165–7195 (2022)
4. Laghrib, A., Afraites, L.: Image denoising based on a variable spatially exponent PDE. Appl. Computa. Harmon. Anal. **68**, 101608 (2024)
5. Nocedal, J., Wright, S.J.: Quadratic programming. Numer. Optim. 448–492 (2006)
6. Afraites, L., Hadri, A., Laghrib, A., Nachaoui, M.: A weighted parameter identification PDE-constrained optimization for inverse image denoising problem. Vis. Comput. **38**(8), 2883–2898 (2022)
7. Hadri, A., Laghrib, A., Nachaoui, A., Nachaoui, M.: On the resolution of the non-smooth inverse Cauchy problem by the primal-dual method. Numer. Algorithms 1–34 (2024)
8. Berdawood, K.A., Nachaoui, A., Nachaoui, M.: An accelerated alternating iterative algorithm for data completion problems connected with Helmholtz equation. Stat. Optim. Inf. Comput. **11**(1), 2–21 (2023)
9. Hadri, A., Nachaoui, M., Laghrib, A., Chakib, A., Afraites, L.: A primal-dual approach for the robin inverse problem in a nonlinear elliptic equation: the case of the $l_1 - l_2$ cost functional. J. Inverse Ill-Posed Probl. **31**(2), 243–266 (2023)
10. Nachaoui, M., Laghrib, A., Hakim, M.: A new space-variant optimization approach for image segmentation. In: Mathematical Control and Numerical Applications, pp. 87–97. Springer, Cham (2021)
11. Jauberteau, F., Nachaoui, M., Zaroual, S.: A novel identification scheme of an inverse source problem based on Hilbert reproducing kernels. In: New Trends of Mathematical Inverse Problems and Applications, pp. 157–175. Springer, Cham (2023)
12. Nachaoui, A., Aboud, F., Nachaoui, M.: Acceleration of the KMF algorithm convergence to solve the Cauchy problem for Poisson's equation. In: Mathematical Control and Numerical Applications: JANO13, Khouribga, Morocco, February 22–24, 2021, pp. 43–57. Springer (2021)
13. Nachaoui, A., Nachaoui, M., Tadumadze, T.: Meshless methods to noninvasively calculate neurocortical potentials from potentials measured at the scalp surface. In: New Trends of Mathematical Inverse Problems and Applications, pp. 99–117. Springer, Cham (2023)

14. Haber, E., Ascher, U.M., Oldenburg, D.: On optimization techniques for solving nonlinear inverse problems. Inverse Probl. **16**(5), 1263 (2000)
15. Berdawood, K., Nachaoui, A., Saeed, R., Nachaoui, M., Aboud, F.: An efficient DN alternating algorithm for solving an inverse problem for Helmholtz equation. Discrete Contin. Dyn. Syst. Ser. S **15**(1) (2022)
16. Nachaoui, M., Nachaoui, A., Tadumadze, T.: On the numerical approximation of some inverse problems governed by nonlinear delay differential equation. RAIRO-Oper. Res. **56**(3), 1553–1569 (2022)
17. Nachaoui, M., Chakib, A., Nachaoui, A.: An efficient evolutionary algorithm for a shape optimization problem. Appl. Comput. Math. **19**(2), 220–244 (2020)
18. Nachaoui, M., Afraites, L., Laghrib, A.: A regularization by denoising super-resolution method based on genetic algorithms. Signal Process. Image Commun. **99**, 116505 (2021)
19. Shanno, D.F.: Conditioning of quasi-newton methods for function minimization. Math. Comput. **24**(111), 647–656 (1970)
20. Mannel, F., Aggrawal, H.O., Modersitzki, J.: A structured L-BFGS method and its application to inverse problems. Inverse Probl. **40**(4), 045022 (2024)
21. Ruder, S.: An Overview of Gradient Descent Optimization Algorithms (2016). arXiv:1609.04747
22. Shamir, O., Zhang, T.: Stochastic gradient descent for non-smooth optimization: convergence results and optimal averaging schemes. In: International Conference on Machine Learning, pp. 71–79. PMLR (2013)
23. Khirirat, S., Feyzmahdavian, H.R., Johansson, M.: Mini-batch gradient descent: faster convergence under data sparsity. In: 2017 IEEE 56th Annual Conference on Decision and Control (CDC), pp. 2880–2887. IEEE (2017)
24. Broyden, C.G.: Quasi-newton methods and their application to function minimisation. Math. Comput. **21**(99), 368–381 (1967)
25. Wright, S.J.: Numerical Optimization (2006)
26. Dai, Y.-H.: Convergence properties of the BFGS algorithm. SIAM J. Optim. **13**(3), 693–701 (2002)
27. Sherman, J., Morrison, W.J.: Adjustment of an inverse matrix corresponding to a change in one element of a given matrix. Ann. Math. Stat. **21**(1), 124–127 (1950)
28. Rondepierre, A.: Méthodes numériques pour l'optimisation non linéaire déterministe. INSA de Toulouse (2017)
29. Yang, Y.: A robust BFGS algorithm for unconstrained nonlinear optimization problems. Optimization **73**(3), 851–873 (2024)
30. Li, D., Wang, X., Huang, J.: Diagonal BFGS updates and applications to the limited memory BFGS method. Comput. Optim. Appl. **81**(3), 829–856 (2022)
31. Byrd, R.H., Nocedal, J., Schnabel, R.B.: Representations of quasi-newton matrices and their use in limited memory methods. Math. Program. **63**(1), 129–156 (1994)
32. Byrd, R.H., Hansen, S.L., Nocedal, J., Singer, Y.: A stochastic quasi-newton method for large-scale optimization. SIAM J. Optim. **26**(2), 1008–1031 (2016)
33. Mannel, F.: A Globalization of L-BFGS for Nonconvex Unconstrained Optimization (2024). arXiv:2401.03805
34. Barzilai, J., Borwein, J.M.: Two-point step size gradient methods. IMA J. Numer. Anal. **8**(1), 141–148 (1988)
35. Li, D.-H., Fukushima, M.: On the global convergence of the BFGS method for nonconvex unconstrained optimization problems. SIAM J. Optim. **11**(4), 1054–1064 (2001)

A Time-Fractional Approach to High-Order PDE Systems in Image Denoising

Ziad Zaabouli, Ayoub Mohssine, Lekbir Afraites, and Amine Laghrib

Abstract This paper introduces a novel time-fractional PDE model for image denoising, which leverages the Caputo fractional derivative to provide enhanced accuracy in preserving fine details during denoising. Our model combines the strengths of Perona-Malik anisotropic diffusion and Weickert's structure tensor, with a time-fractional evolution equation to account for non-local temporal effects. The model is formulated as a reaction-diffusion system, and we establish the existence and uniqueness of solutions using the Schauder fixed point theorem. Numerical results show that our method improves texture preservation while effectively removing noise, outperforming existing methods in both visual quality and computational efficiency.

Keywords Image denoising · Coupled PDE · Fractional calculus · Image decomposition · Schauder fixed point

1 Introduction

Image denoising remains a fundamental challenge in signal processing and computer vision [1–4], requiring mathematical models that can effectively reduce noise while preserving essential features such as edges, fine details, and textures. The goal is to restore a clean image U from noisy observations f, which can be formulated as follows:

Lekbir Afraites and Amine Laghrib are contributed equally to this work.

Z. Zaabouli · A. Mohssine · L. Afraites
EMI, Sultan Moulay Slimane University, Beni-Mella, Morocco
e-mail: ayoub.mohssine@edu.uca.ac.ma

L. Afraites
e-mail: l.afraites@usms.ma

A. Laghrib (✉)
ENSA Khouribga, EMI, Université Sultan Moulay Slimane, Beni-Mella, Morocco
e-mail: laghrib.amine@gmail.com

A. Laghrib et al. (eds.), *Computational Methods for Inverse Problems and Applications*, Springer Proceedings in Mathematics & Statistics 498,
https://doi.org/10.1007/978-3-031-89498-5_14

$$f = U + N, \tag{1}$$

where N is the noise component. Given the ill-posed nature of this problem, a common approach is to incorporate prior knowledge into the model, leading to the following minimization problem:

$$\min_U E(U) = F(U, f) + \lambda R(U), \tag{2}$$

where F represents the data fidelity term, R is the regularization term, and $\lambda > 0$ is a regularization parameter that balances these terms. Over the years, various regularization techniques have been proposed in the literature to address this minimization problem [5, 6]. These techniques typically involve variational methods and PDE-based models [7–10]. Among the most influential is the total variation (TV) minimization introduced by Rudin et al. [11], which aims to preserve edges while reducing noise. The associated energy functional is:

$$\min_U E(U) = \lambda \|U - f\|^2_{L^2(\Omega)} + \int_\Omega |\nabla U|, \tag{3}$$

where Ω is an open and bounded domain in $\mathbb{R}^2$, and $f \in L^2(\Omega)$. The associated time-dependent partial differential equation derived from the Euler-Lagrange equation of $E(u)$ is given by:

$$\left\{ \frac{\partial U}{\partial t} = \operatorname{div}\left(\frac{\nabla U}{|\nabla U|}\right) + 2\lambda(f - U) \quad \text{in } \Omega, \ \langle \nabla U, n \rangle = 0 \quad \text{on } \partial\Omega. \right. \tag{4}$$

The TV-based model (4) has proven effective for image denoising and edge preservation. However, it may overlook minor features such as textures (see [12]). To address these shortcomings, Meyer [13] introduced a new minimization method that uses a weaker norm to better capture oscillatory patterns and fine image details. This led to the development of higher-order PDE models, such as those incorporating the H^{-1} norm, as proposed by Osher et al. [12]. The corresponding PDE-based functional is:

$$\| \cdot \|_{H^{-1}(\Omega)} = \int_\Omega |\nabla \Delta^{-1}(\cdot)|^2, dx. \tag{5}$$

Leading to a fourth-order PDE:

$$\left\{ \frac{\partial U}{\partial t} = \Delta\left(\operatorname{div}\left(\frac{\nabla U}{|\nabla U|}\right)\right) + 2\lambda(f - U) \quad \text{in } \Omega, \ \langle \nabla U, n \rangle = 0 \quad \text{on } \partial\Omega, \right. \tag{6}$$

Although fourth-order PDEs improve texture preservation, their numerical complexity has led researchers to explore alternative approaches. To overcome these computational challenges, researchers have reformulated fourth-order models into systems of coupled second-order equations. These systems not only simplify the numerical implementation but also offer improved stability and accuracy. For example,

Guo et al. [14] proposed a decomposition technique where the image f is split into a cartoon part U (representing the smoothed image) and a texture or noise part V, resulting in the following coupled system:

$$\begin{cases} \frac{\partial U}{\partial t} = \operatorname{div}\left(\frac{\nabla U}{|\nabla U|}\right) - 2\lambda V & \text{in } \Omega \times (0, T), \\ \frac{\partial V}{\partial t} = \Delta V - (f - U) & \text{in } \Omega \times (0, T), \\ \langle \nabla U, n \rangle = \langle \nabla V, n \rangle = 0 & \text{on } \partial\Omega \times (0, T), \\ u(0, x) = f & \text{in } \Omega, \\ v(0, x) = 0 & \text{in } \Omega. \end{cases} \tag{7}$$

Recently, researchers have shown considerable interest in various adaptations of this model. For detailed insights, we point readers to comprehensive studies [15–17] that exhibit promising results.

As the theory of fractional calculus continues to evolve, its application has expanded across various fields, including physics, engineering, and, more recently, image processing [18, 19]. Fractional derivatives, which involve non-integer orders, offer a powerful framework for capturing the intricate details of systems where classical integer-order models fail to fully characterize the underlying complexity. Unlike traditional derivatives that represent local changes, fractional derivatives operate on a nonlocal scale, incorporating the history of a function's past states to model dynamic systems more accurately. This nonlocal behavior makes fractional derivatives particularly suited for problems involving memory effects or long-range interactions. Among the various definitions of fractional derivatives, the Caputo derivative [20, 21] has emerged as one of the most widely used in applied sciences due to its practical interpretation of initial conditions. This derivative is particularly advantageous in the context of physical problems as it allows for the use of classical initial conditions, making it more intuitive in real-world applications. The Caputo derivative, is defined for providing a continuous transition between classical integer-order differentiation and fractional calculus. In the context of image processing, the Caputo fractional derivative is valuable for enhancing algorithms that require the incorporation of memory effects and nonlocal dependency, two features that are critical in tasks like image denoising, super-resolution, and texture recovery. The Caputo derivative captures how the evolution of the image over time is influenced by both current and previous states, effectively modeling the cumulative impact of the image's history. This makes it particularly well-suited for applications like image denoising, where preserving fine details such as edges and textures while reducing noise is essential. By leveraging the fractional time-derivative, denoising models can take into account long-term dependencies in the pixel intensity changes, thus improving the denoising process while minimizing the loss of crucial image features.

Building upon the above discussion, which highlights the advantages of the decomposition strategy using the H^{-1} norm, the Weickert filter [22], and the Caputo time-fractional derivative, we propose a novel image denoising model that combines these powerful mathematical tools into a unified framework. The decomposition strategy of the H^{-1} norm enables the separation of the image into a smooth component and a texture/noise component, ensuring better preservation of fine details and textures during the denoising process. The Weickert anisotropic diffusion filter contributes by enhancing edge preservation, directing diffusion along the structure of the image, thereby preventing excessive smoothing near critical regions like contours and corners. Finally, by integrating the Caputo fractional derivative, the model incorporates memory effects, capturing the temporal evolution of the image in a non-local manner. This combination results in a highly effective denoising framework that excels in maintaining the balance between noise reduction and the preservation of important image features.

Our contributions in this paper are outlined as follows: In Sect. 2, we introduce the our new model. Section 3 provides the well-posedness analysis of the system. In Sect. 4, we detail the discretization scheme of the model. Finally, Sect. 5 presents numerical experiments and comparisons, demonstrating the effectiveness and performance of the proposed model.

2 The Proposed Reaction-Diffusion System

In this section, we introduce our new model that combines the Weickert anisotropic diffusion filter [22] with the Caputo time fractional derivative to enhance both edge preservation and memory effects in the image denoising process. The proposed system is formulated as follows:

$$\begin{cases} \partial^s_{0,t} U - div\big(\mathfrak{D}(\mathbb{J}_\rho(\nabla U_\sigma))\nabla U\big) + \lambda V = 0 & \text{in } Q := (0,T)\times\Omega \\ \partial^s_{0,t} V - \beta\Delta V + (f-U) = 0 & \text{in } Q := (0,T)\times\Omega \\ \langle \mathfrak{D}(\mathbb{J}_\rho(\nabla U_\sigma))\nabla U, n\rangle = \langle \nabla V, n\rangle = 0 & \text{in } \Sigma := (0,T)\times\partial\Omega \\ U(0,x) = f, \quad V(0,x) = 0 & \text{in } \Omega, \end{cases} \tag{8}$$

where $\Omega \subset \mathbb{R}^2$ is a bounded smooth domain with Lipschitz continuous boundary $\partial\Omega$, $\beta > 0$.

For every $s \in (0,1)$ the Caputo time fractional derivative is defined as follows

$$\partial^s_{0,t} U(t,x) := \frac{1}{\Gamma(1-s)}\int_0^t (t-\tau)^{-s}\frac{\partial U(\tau,x)}{\partial \tau}d\tau, \text{ a.e. } \tau \in [0,T].$$

The $\mathfrak{D}$ is an anisotropic diffusion tensor and $\mathbb{J}_\rho$ is the structure tensor defined by

$$\mathbb{J}_\rho(\nabla U_\sigma) = K_\rho * (\nabla U_\sigma \otimes \nabla U_\sigma). \tag{9}$$

The symbol $*$ is the convolution product and $\otimes$ is the outer product defined such as

$$\nabla U_\sigma \otimes \nabla U_\sigma = \nabla U_\sigma (\nabla U_\sigma)^t.$$

Here U_σ is constructed using a convolution of U with a Gaussian kernel, while K_ρ and K_σ are Gaussian convolution kernels given by $K_\tau(x) = \frac{1}{2\pi\tau^2}\exp(-\frac{|x|^2}{2\tau^2})$. The tensor $\mathfrak{D}$ is computed using the eigenvalues and eigenvectors of $\mathbb{J}_\rho$ as follows

$$\mathfrak{D} := g_+(\lambda_1, \lambda_2)\theta_1\theta_1^\intercal + g_-(\lambda_1, \lambda_2)\theta_2\theta_2^\intercal = (\theta_1 \;\; \theta_2)\begin{pmatrix} g_+(\lambda_1, \lambda_2) & 0 \\ 0 & g_-(\lambda_1, \lambda_2) \end{pmatrix}(\theta_1 \;\; \theta_2)^t \quad (10)$$

where θ_1, θ_1 are the eigenvectors associated to the eigenvalues λ_1, λ_2 of the tensor structure $\mathbb{J}_\rho$. While the functions g_+ and g_- describe the isotropic or anisotropic behavior of the smoothing on the image regions.

Before proving the well-posedness of the solution to problem (8), we impose the following assumptions (H):

H_1 The tensors $\mathfrak{D}$ belongs to $C^\infty(\mathbb{R}^{2\times 2}, \mathbb{R}^{2\times 2})$, is coercive with coercivity constant δ and positive-definite matrix.

H_2 The function f is in $L^2(\Omega)$.

H_3 The constants σ, β, λ, s, ρ, and δ are all positive.

3 Main Results

The solution to the proposed denoising model is defined in the following sense:

Definition 1 A pair of measurable functions (U, V) is said to be a weak solution to fractional problem (8) if they satisfies

$$(U, V) \in \left(L^2\left(0, T; H^1(\Omega)\right)\right)^2, (\partial^s_{0,t}U, \partial^s_{0,t}V) \in \left(L^2\left(0, T; (H^1(\Omega))'\right)\right)^2$$

and

$$\begin{aligned} \langle \partial^s_{0,t}U, \Phi\rangle + \int_\Omega \mathfrak{D}(\mathbb{J}_\rho(\nabla U_\sigma))\nabla U \nabla\Phi dx + \lambda \int_\Omega V\Phi dx = 0, \\ \langle \partial^s_{0,t}V, \Phi\rangle + \frac{\beta}{2}\int_\Omega \nabla V \nabla \Phi dx + \int_\Omega (f - U)\Phi dx = 0, \end{aligned} \quad (11)$$

for every $\Phi \in H^1(\Omega)$.

To demonstrate the existence of a solution to problem (8), we first require the following a priori estimates.

Theorem 1 *Under the hypothesis (H), the solution of the problem (11) satisfies the following estimations:*

$$\begin{aligned} \|U\|_{L^2(0,T;H^1(\Omega))} + \|U\|_{L^\infty(0,T;L^2(\Omega))} + \|\partial_t^s U\|_{L^2(0,T;H^1(\Omega)')} \leq C, \\ \|V\|_{L^2(0,T;W^{s,2}(\Omega))} + \|V\|_{L^\infty(0,T;L^2(\Omega))} + \|\partial_t^s V\|_{L^2(0,T;H^1(\Omega)')} \leq C, \end{aligned} \tag{12}$$

C is a constant that depends only on initial data.

Next, we establish the existence and uniqueness of a weak solution to the system 8.

Theorem 2 *Under the assumptions (H) above, the problem (8) have a unique weak solution* $(U, V) \in C(0, T; L^2(\Omega)) \cap L^2(0, T; H^1(\Omega))$.

Remark 1 The proofs of Lemma (1) and Theorem (1) follow similar arguments as those provided in [23].

4 Discrete Setting

This section is devoted to the numerical approximation of the proposed model, involving the discretization of problem (8) through a finite difference scheme. To do so, we start by defining k as the time step size and h the spatial grid size, w resulting in the following time and space discretization:

$$\begin{aligned} t_n &= nk, \quad n = 0, 1, 2, \ldots, \\ x_i &= ih, \quad i = 0, 1, 2, \ldots, M, \\ y_j &= jh, \quad j = 0, 1, 2, \ldots, N, \end{aligned}$$

where $N \times M$ represent the size of the image domain Ω. Let $f_{i,j}$, $V_{i,j}^n$ and $U_{i,j}^n$ represent the estimations of $f(x_i, y_j)$, $V(t_n, x_i, y_j)$ and $U(t_n, x_i, y_j)$, respectively.

Our primary objective is to approximate the Caputo time fractional derivative, and we accomplish this by following the method suggested by [24] to address the time derivative. This approximation can be expressed as:

$$\partial_{0,t}^s U(t_{n+1}, x_i, y_j) \simeq \kappa_{\alpha,k} \sum_{m=1}^{n+1} \omega_m^\alpha (U_{i,j}^{n-m+2} - U_{i,j}^{n+1-m}) \tag{13}$$

$$= \kappa_{\alpha,k} \Big(U_{i,j}^{n+1} - \sum_{m=1}^{n} (\omega_m^\alpha - \omega_{m+1}^\alpha) U_{i,j}^{n+1-m} - \omega_{n+1}^\alpha U_{i,j}^0 \Big), \tag{14}$$

where $\kappa_{\alpha,k} = \dfrac{1}{k^\alpha \Gamma(1-\alpha)(1-\alpha)}$ and $\omega_m^\alpha = m^{1-\alpha} - (m-1)^{1-\alpha}$.

Our next step is to discretized the diffusion term $\mathrm{div}(\mathfrak{D}(\mathbb{J}_\rho(\nabla U_\sigma))\nabla U)$. Numerous schemes have been proposed in the literature to discretize this term. Here, we consider a classical technique introduced by Weickert in [22]. Next, we will denote

$$\mathfrak{D}(\mathbb{J}_\rho(\nabla U_\sigma)) = \begin{pmatrix} a & b \\ b & c \end{pmatrix}, \tag{15}$$

with

$$\begin{aligned} a &= \mu_1 \cos^2\alpha + \mu_2 \sin^2\alpha, \\ b &= (\mu_1 - \mu_2)\sin\alpha\cos\alpha, \\ c &= \mu_1 \sin^2\alpha + \mu_2 \cos^2\alpha. \end{aligned}$$

where α is giving by

$$\tan(2\alpha) = \frac{2\mathbb{J}_{12}}{\mathbb{J}_{11} - \mathbb{J}_{22}},$$

with $\mathbb{J}_{12}$, $\mathbb{J}_{11}$ and $\mathbb{J}_{22}$ are the components of the structure tensor $\mathbb{J}_\rho$, that are

$$\mathbb{J}_\rho(\nabla U_\sigma) = \begin{pmatrix} \mathbb{J}_{11} & \mathbb{J}_{12} \\ \mathbb{J}_{12} & \mathbb{J}_{22} \end{pmatrix},$$

and μ_1, μ_2 the corresponding eigenvalues of $\mathbb{J}_\rho$.

So, the anisotropic diffusion term $\mathrm{div}(\mathfrak{D}(\mathbb{J}_\rho(\nabla u_\sigma))\nabla U^n)$ can be developed as

$$\begin{aligned} \mathrm{div}(\mathfrak{D}(\mathbb{J}_\rho(\nabla U_\sigma))\nabla U^n) &= \mathrm{div}\left[\begin{pmatrix} a & b \\ b & c \end{pmatrix}\begin{pmatrix} \partial_x U^n \\ \partial_y U^n \end{pmatrix}\right] \\ &= \partial_x\left(a\partial_x U^n\right) + \partial_x\left(b\partial_y U^n\right) + \partial_y\left(b\partial_x U^n\right) + \partial_y\left(c\partial_y U^n\right), \end{aligned}$$

where

$$\begin{aligned} \partial_x\left(a\partial_x U^n\right) &= \partial_x a \partial_x U^n + a\partial_{xx} U^n, \\ \partial_x\left(b\partial_y U^n\right) &= \partial_x b \partial_y U^n + b\partial_{xy} U^n, \\ \partial_y\left(b\partial_x U^n\right) &= \partial_y b \partial_x U^n + b\partial_{yx} U^n, \\ \partial_y\left(c\partial_y U^n\right) &= \partial_y c \partial_y U^n + c\partial_{yy} U^n. \end{aligned}$$

We denote by $(\partial_x U^n)_{i,j}$, $(\partial_y U^n)_{i,j}$, $(\partial_{xx} U^n)_{i,j}$, $(\partial_{xy} U^n)_{i,j}$, $(\partial_{yy} U^n)_{i,j}$ the discrete version respectively of the operators $\partial_x U^n$, $\partial_y U^n$, $\partial_{xx} U^n$, $\partial_{xy} U^n$, $\partial_{yy} U^n$, that are

$$\left(\partial_x U^n\right)_{i,j} = \begin{cases} u^n_{i+1,j} - u^n_{i,j} & \text{if } i < N \\ 0 & \text{if } i = N \end{cases}, \quad \left(\partial_y U^n\right)_{i,j} = \begin{cases} u^n_{i,j+1} - u^n_{i,j} & \text{if } j < M \\ 0 & \text{if } j = M \end{cases}. \tag{16}$$

The second derivatives operators are approximated as

$$(\partial_{xx}U^n)_{i,j} = \begin{cases} u^n_{i+1,j} - 2u^n_{i,j} + u^n_{i-1,j} & \text{si } 1 < i < N, \\ u^n_{i+1,j} - u^n_{i,j} & \text{si } i = 1, \\ u^n_{i-1,j} - u^n_{i,j} & \text{si } i = N \end{cases}$$

$$\left(\partial_{xy}U^n\right)_{i,j} = \begin{cases} u^n_{i,j+1} - u^n_{i,j} - u^n_{i-1,j+1} + u^n_{i-,j} & \text{si } 1 < i < N, 1 < j < M \\ 0 & \text{si } i = 1, \\ 0 & \text{si } i = N \end{cases}$$

$$\left(\partial_{yx}U^n\right)_{i,j} = \begin{cases} u^n_{i+1,j} - u^n_{i,j} - u^n_{i+1,j-1} + u^n_{i,j-1} & \text{si } 1 < i < N, 1 < j < M \\ 0 & \text{si } j = 1 \\ 0 & \text{si } j = M \end{cases}$$

$$\left(\partial_{yy}U^n\right)_{i,j} = \begin{cases} u^n_{i,j+1} - 2u^n_{i,j} + u^n_{i,j-1} & \text{si } 1 < j < M \\ u^n_{i,j+1} - u^n_{i,j} & \text{si } j = 1, \\ u^n_{i,j-1} - u^n_{i,j} & \text{si } j = M \end{cases}$$

Utilizing these notations, we can formulate the discretization of the diffusion term as:

$$\left[\text{div}\left(\mathfrak{D}(\mathbb{J}_\rho(\nabla u_\sigma))\nabla U^n\right)\right]_{i,j} = (\partial_x a)_{i,j}(\partial_x U^n)_{i,j} + (a)_{i,j}(\partial_{xx}U^n)_{i,j} + (\partial_x b)_{i,j}(\partial_y U^n)_{i,j} \tag{17}$$

$$+ (b)_{i,j}(\partial_{xy}U^n)_{i,j} + (\partial_y b)_{i,j}(\partial_x U^n)_{i,j} + (b)_{i,j}(\partial_{yx}U^n)_{i,j} \tag{18}$$

$$+ (\partial_y c)_{i,j}(\partial_y U^n)_{i,j} + (c)_{i,j}(\partial_{yy}U^n)_{i,j}. \tag{19}$$

Using these notations, we can express the discretization of the diffusion term as follows: Specifically, the standard discretization of (17) involves convolving the discrete image u at each point (i, j) with a 3×3 Stencil matrix denoted as S. Additionally, one of the conditions imposed by Weickert [22] on the matrix associated to the diffusion term is the positivity of its elements, which is required for the stability of this schema. Let $\mathfrak{M}(U^n)$ indicates the matrix containing the elements of term $\text{div}(\mathfrak{D}(\mathbb{J}_\rho(\nabla u_\sigma))\nabla U^n)$ and S the Stencil matrix giving as:

$$S = \begin{pmatrix} s_{11} & s_{12} & s_{13} \\ s_{21} & s_{22} & s_{23} \\ s_{31} & s_{32} & s_{33} \end{pmatrix}.$$

then, an element of $\mathfrak{M}(U^n)$ is calculated by:

$$\begin{aligned}[\mathfrak{M}(U^n)]_{i,j} &= u^n_{i-1,j-1}s_{11} + u^n_{i-1,j}s_{12} + u^n_{i-1,j+1}s_{13} \\ &+ u^n_{i,j-1}s_{21} + u^n_{i,j}s_{22} + u^n_{i,j+1}s_{23} \\ &+ u^n_{i+1,j-1}s_{31} + u^n_{i+1,j}s_{32} + u^n_{i+1,j+1}s_{33}.\end{aligned}$$

Finally, the discrete explicit scheme of the problem (8) could be written as

$$\begin{cases} U_{i,j}^{n+1} = \sum_{m=1}^{n}(\omega_m^\alpha - \omega_{m+1}^\alpha)U_{i,j}^{n+1-m} - \omega_{n+1}^\alpha U_{i,j}^0 + \frac{1}{\kappa_{\alpha,k}}([\mathfrak{M}(U^n)]_{i,j} - \lambda V_{i,j}^n), \\ V_{i,j}^{n+1} = \sum_{m=1}^{n}(\omega_m^\alpha - \omega_{m+1}^\alpha)V_{i,j}^{n+1-m} - \omega_{n+1}^\alpha V_{i,j}^0 \\ +\frac{1}{\kappa_{\alpha,k}}\left(\beta\frac{V_{i,j-1}^n + V_{i,j+1}^n - 4V_{i,j}^n + V_{i-1,j}^n + V_{i+1,j}^n}{h^2} - f_{i,j} + U_{i,j}^n\right), \\ U_{i,j}^0 = f_{i,j},\ V_{i,j}^0 = 0. \end{cases} \tag{20}$$

for all $n \geq 1, 0 \leq i \leq N$ and $0 \leq j \leq M$. The main coupled algorithm associated to the problem (8) is summarized in Algorithm 1.

Data: Given a noisy image f and an initial $V^0 = 0$. Choose $\delta, \beta, \lambda, s \in (0, 1), \rho, \sigma, T > 0, \epsilon$.

Result: The restored image U and the Cartoon part V

1. **Repeat**
2. Compute U^{n+1} by solving the first equation of the system (20).
3. Compute V^{n+1} by solving the second equation of the system (20).
4. **Until** $\frac{\|U^{n+1}(T)-U^n(T)\|_{L^2}}{\|U^n(T)\|_{L^2}} \leq \epsilon$.

Algorithm 1: Coupled PDE system algorithm

5 Numerical Results

In this section, we analyze the effectiveness of the proposed denoising model in restoring clean images that have been corrupted by Impulse noise at varying levels. For our algorithm parameters, we initialize: $\delta = 4 \times 10^{-2}$, $\beta = 0.1$, $\lambda = 0.02$, $\rho = 3.2$, $\sigma = 35$, $\epsilon = 10^{-4}$ and $s = 0.6$. On average, the primary algorithm, implemented in MATLAB R2015, requires approximately 2–3 min to process conventional 256×256 grayscale images on an i7 3.4 GHz computer. Processing color images significantly increases computation time, prompting the use of parallel computation.

We examine four test images with distinct characteristics and varying levels of Gaussian noise, with parameters $r = 0.02$, $r = 0.03$, $r = 0.04$, and $r = 0.05$. The evaluation focuses on the proposed model's capability to maintain image details and textures while effectively reducing noise. Specifically, we analyze how the model preserves sharp edges and structural patterns, particularly in regions with pronounced anisotropy. The comparison is done with Meyer Model [13], Guo et al. equation [14]. Figure 1 illustrate the denoised outputs using our model and compared to other denoising approaches. The results across all tests indicate the model's ability to

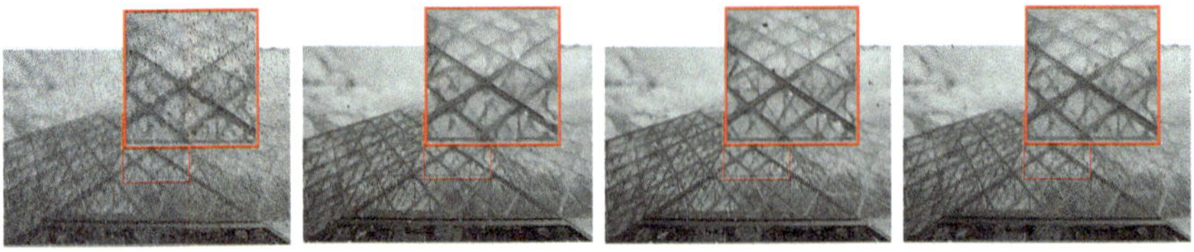

(a) Noisy image (b) Meyer Model [13] (c) Guo et al. [14] (d) Our

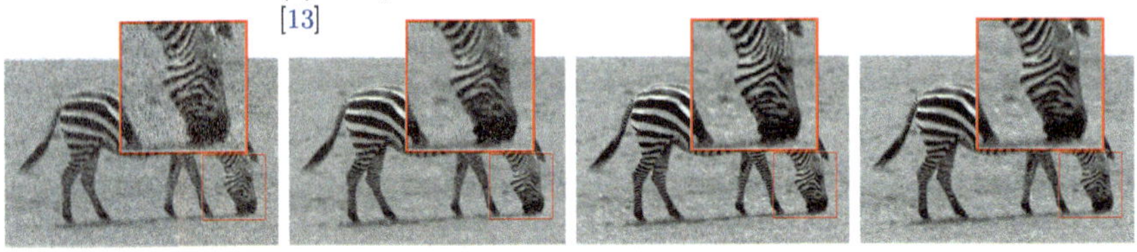

(e) Noisy image (f) Meyer Model [13] (g) Guo et al. [14] (h) Our

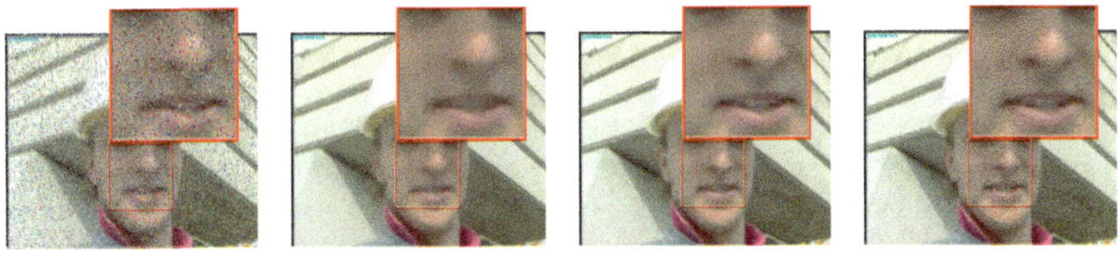

(i) Noisy image (j) Meyer Model [13] (k) Guo et al. [14] (l) Our

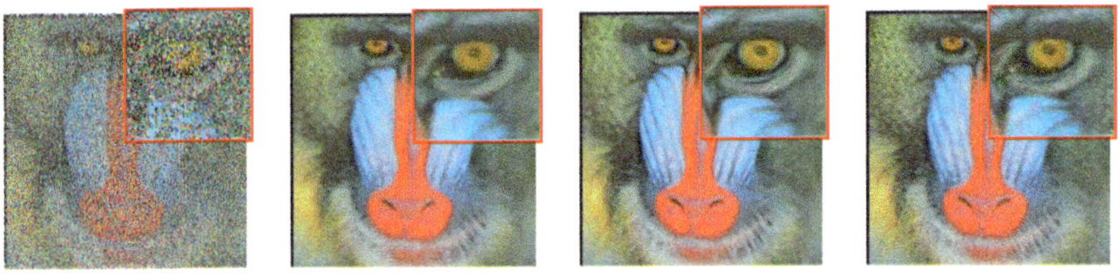

(m) Noisy image (n) Meyer Model [13] (o) Guo et al. [14] (p) Our

Fig. 1 Reconstruction of natural tests: *Pond*, *Zebra*, *Man* and *Baboon*

effectively preserve edges and textures while significantly reducing noise, even when applied to images with diverse characteristics (synthetic, geometric, or textured).

In Table 1, we provide a quantitative comparison of the four examples using two critical metrics: the Peak Signal-to-Noise Ratio (PSNR), which assesses the strength of the signal relative to the noise in the image, and the Structural Similarity Index (SSIM), which evaluates the similarity between two images. Notably, the proposed method consistently outperformed the competing approaches across all evaluated cases.

Table 1 The PSNR and SSIM values for the evaluated methods on four test images

Image	Method	PSNR	SSIM	Image	Method	PSNR	SSIM
Pond	Meyer model [13]	30.65	0.901	*Zebra*	Meyer model [13]	28.83	0.800
	Guo et al. [14]	31.01	0.902		Guo et al. [14]	28.72	0.822
	Our	32.08	0.923		Our	28.99	0.831
Man	Meyer model [13]	27.25	0.725	*Baboon*	Meyer model [13]	27.48	0.780
	Guo et al. [14]	27.85	0.774		Guo et al. [14]	28.59	0.794
	Our	28.47	0.765		Our	28.82	0.811

6 Conclusion

In conclusion, we have introduced and rigorously analyzed a novel image denoising model that integrates advanced techniques from fractional calculus, anisotropic diffusion, and the H^{-1} decomposition strategy. Our model employs the Weickert anisotropic filter in conjunction with a Caputo time-fractional derivative to address the challenge of removing noise while preserving critical image features such as edges and textures. Through theoretical analysis, we have established the well-posedness of the model and provided numerical experiments that demonstrate its effectiveness. The results indicate that our model achieves superior denoising performance compared to traditional methods, making it a valuable tool for practical image processing applications.

Conflict of Interest The authors have no conflict of interest.

References

1. Laghrib, A., Afraites, L.: Image denoising based on a variable spatially exponent PDE. Appl. Comput. Harmon. Anal. **68**, 101608 (2024)
2. Laghrib, A., Afraites, L., Nachaoui, M.: New Trends of Mathematical Inverse Problems and Applications
3. Laghrib, A., Limami, F.-E., Nachaoui, A.: Enhanced osmosis model with bilateral total variation for effective shadow removal. Comput. Math. Appl. **174**, 71–83 (2024)
4. Laghrib, A., Bella, F.Z.A., Nachaoui, M., Jauberteau, F.: Bilevel optimal parameter learning for a high-order nonlocal multiframe super-resolution problem. Inverse Probl. **40**(1), 015004 (2023)
5. Nachaoui, M., Laghrib, A., Hakim, M.: A new space-variant optimization approach for image segmentation. In: International Conference on Numerical Analysis and Optimization Days, pp. 87–97. Springer (2021)
6. Nachaoui, M., Afraites, L., Laghrib, A.: A regularization by denoising super-resolution method based on genetic algorithms. Signal Process. Image Commun. **99**, 116505 (2021)
7. Afraites, L., Hadri, A., Laghrib, A.: A denoising model adapted for impulse and Gaussian noises using a constrained-PDE. Inverse Probl. **36**(2), 025006 (2020)
8. Afraites, L., Hadri, A., Laghrib, A., Nachaoui, M.: A high order PDE-constrained optimization for the image denoising problem. Inverse Probl. Sci. Eng. **29**(12), 1821–1863 (2021)

9. Hadri, A., Afraites, L., Laghrib, A., Nachaoui, M.: A novel image denoising approach based on a non-convex constrained PDE: application to ultrasound images. Signal Image Video Process. **15**, 1057–1064 (2021)
10. El Hakoume, A., Zaabouli, Z., Laghrib, A., Afaites, L.: Image restoration using a coupled reaction-diffusion equations. In: New Trends of Mathematical Inverse Problems and Applications, pp. 137–155. Springer, Cham (2023)
11. Rudin, L.I., Osher, S., Fatemi, E.: Nonlinear total variation based noise removal algorithms. Phys. D: Nonlinear Phenom. **60**(1–4), 259–268 (1992)
12. Osher, S., Solé, A., Vese, L.: Image decomposition and restoration using total variation minimization and the h^{-1} norm. Multiscale Model. Simul. **1**(3), 349–370 (2003)
13. Meyer, Y.: Oscillating Patterns in Image Processing and Nonlinear Evolution Equations: The Fifteenth Dean Jacqueline B. Lewis Memorial Lectures, vol. 22. American Mathematical Soc. (2001)
14. Guo, Z., Liu, Q., Sun, J., Wu, B.: Reaction-diffusion systems with p(x)-growth for image denoising. Nonlinear Anal. Real World Appl. **12**(5), 2904–2918 (2011)
15. Hakoume, A.E., Zaabouli, Z., Afraites, L., Laghrib, A.: On a mathematical analysis of a coupled system adapted to MRI image denoising. J. Nonlinear Sci. **33**(6), 113 (2023)
16. Zaabouli, Z., Afraites, L., Laghrib, A., Hadri, A.: A variable p[u] exponent reaction-diffusion PDE for image denoising. J. Math. Anal. Appl. 128624 (2024)
17. Zaabouli, Z., El Hakoume, A., Afraites, L., Laghrib, A.: A novel coupled p(x) and fractional PDE denoising model with theoretical results. Int. J. Comput. Math. 1–26 (2024)
18. Oulmelk, A., Srati, M., Afraites, L.: Comparing numerical methods for inverse source problem in time-fractional diffusion equation. In: New Trends of Mathematical Inverse Problems and Applications, pp. 1–16. Springer, Cham (2023)
19. Oulmelk, A., Afraites, L., Hadri, A., Nachaoui, M.: An optimal control approach for determining the source term in fractional diffusion equation by different cost functionals. Appl. Numer. Math. **181**, 647–664 (2022)
20. Kilbas, A.A., Srivastava, H.M., Trujillo, J.J.: Theory and Applications of Fractional Differential Equations, vol. 204. Elsevier (2006)
21. Boukdir, A., Laghrib, A., Nachaoui, M.: Time-fractionated anisotropic osmosis with bilateral total variation for effective shadow mitigation in image processing. Appl. Math. Model. **137**, 115671 (2025)
22. Weickert, J., et al.: Anisotropic Diffusion in Image Processing, vol. 1. Teubner, Stuttgart (1998)
23. El Hakoume, A., Afraites, L., Laghrib, A.: An improved coupled PDE system applied to the inverse image denoising problem. Electron. Res. Arch. **30**(7), 2618–2642 (2022)
24. Murio, D.A.: Implicit finite difference approximation for time fractional diffusion equations. Comput. Math. Appl. **56**(4), 1138–1145 (2008)

The manufacturer's authorised representative in the EU is Springer Nature Customer Service Centre GmbH, Europaplatz 3, 69115 Heidelberg, Germany. If you have any concerns regarding our products, please contact ProductSafety@springernature.com

Printed and bound by CPI Group (UK) Ltd, Croydon, CR0 4YY
30/09/2026
02225471-0001